NEARLY A MINER

The author, aged 22 yrs, in his pit-clothes.

NEARLY A MINER

Einion Evans

First impression—June 1994

ISBN 1 75902 064 X

Printed by
J.D. Lewis & Sons Ltd., Gomer Press, Llandysul, Dyfed.

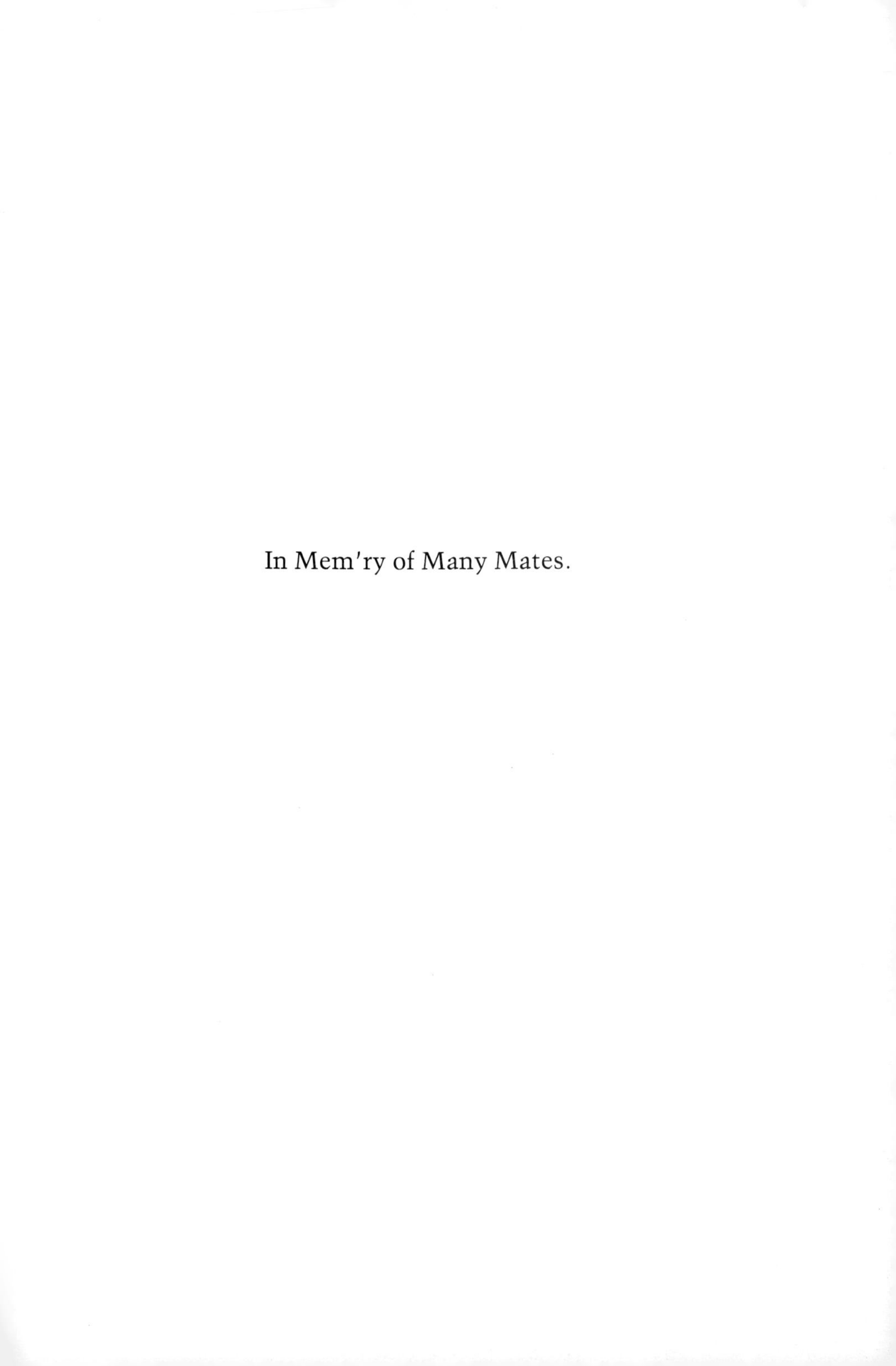

In Mem'ry of Many Mates.

An Account of the Contents

I Think That I Owe These Thanks 9

Around Where I Have My Roots 11

Tall Hopes in a Little Home 17

How I Saw the Picton Scene 23

A Small Snap of Both Chapels 31

In Touch With Yesterday's Team 40

A Landmark Stark and So Deep 45

I Scan Through My Days at School 54

Snappin With the Bevin Boys 66

Courting and Loving For Life 76

Into Heights of Parenthood 86

An Avowed Eisteddfodic 94

Now More About My Heroes 104

Coal-mining and Singing Songs 113

These Were the Ponies Down Pit 122

A Lot in Which I'm Well Pleased 131

To Help With Some of the Terms 138

I Think That I Owe These Thanks

Betty for her continued support.
Norman Closs Parry for taking some of the photographs.
The following for the loan of valuable photographs:
Glenys Brooks Jones, *The County Herald* through Elfed Hughes, Gwenda Davies, John Herbert Jones, Ted McKay, John Morris, Nefyn Morris, Blodwen Owen, Olwen Owens, Jean Glynne Parry, Yvonne Parry, Rhiannon Parry, Bill Unsworth, David Wright.
The Publishers Gomer Press for their usual high standard.
Linda Miller for typing the translation.

Around Where I have My Roots

Picton. Yes, that's it, Picton. I know that only comparatively few people know about the existence of the place by now, and fewer people know of its Welsh name, Butyn. In the Electoral Register, the region, together with nearby villages which are far larger than Picton are referred to as 'The Township of Picton'. Cause for justifiable pride there for the locals of Picton, who have also received a mention in the 'Doomsday Book'. The name given to one of the many seams of coal in the Point of Ayr is 'Bychdyn'. My education is totally inadequate for me to pontificate on place-names, but I have enough between my ears to see a connection between Picton, Butyn, and Bychtyn. Yes indeed, the roots are quite deep.

In order to be deemed a village I believe a place must have a church, a public house, and a school. Seemingly the residents of Picton were unperturbed by having to walk out of their locality for spiritual and temporal blessings, but they insisted on obtaining education for their children, although the school was at one of the extremities of the region—Gwespyr-Picton-Council School. Passers-by are surprised to see a school in such a splendid building, of Gwespyr stone, in a place with so few houses around. From the First World War period up to the fifties children walked perfectly safely to school from Gwespyr, Picton, Glan'rafon, and Penyffordd. Because of its central situation to these places, practically all the children lived within a mile of the school. Notable exceptions to this rule were Talbot Taylor from Talacre Lighthouse, and Tudor, and Trevor Davies from Ty'n-y-Morfa. The dangers from traffic were practically nil, as also was the danger of children and women being molested. I remember how one of the teachers, Miss Parry, from Dyserth, would travel by bus to Tan-lan. She would climb the wooded hill, and then cross Peter Williams's fields before reaching the school. Through the years she was not even frightened, except maybe by an occasional cow that would come too close to the foot-path.

Picton did not possess a church or chapel, although the residents knew how to bend the knee; there was no pub either, although some knew how to bend the elbow. Only five farms, four small-holdings, six houses, and seven single-storey cottages claimed about a mile of Sellafield-free air between Penyffordd and Gwespyr in one direction, and Glan'rafon and Point of Ayr in the other direction. The English names given to the five farms were totally unimaginative—Picton Farm, Upper Picton Farm, Picton Dairy Farm, Picton Hall Farm and Picton Yard Farm. However the names given to some of the fields were much more melodious—Y Weirglodd, Cae Gegin, Cae Dafydd, Cae Caban, Berllin Uchaf, Berllin Isaf, Cae Tom Parry, Maes Mari, Cae Dŵr Coch, Trumiau, Geufron, Fron Las, Aceri Hirion, Cae Bryn, and Cae'r Brêc. The reason for the Anglicised names of the farms can be traced to one of their early owners, Sir Piers Mostyn, of Talacre, because as we all know, in his family, by the beginning of this century, the Welsh language had been diluted to the weakness of a little wren's urine. It was the local tenants who gave the Welsh names to the fields.

The place has only seen a few changes over the years—in my time four new houses and two new bungalows. Some of the houses have been modernised, naturally (or maybe unnaturally!). One of the small-holdings called 'Hyfrydle' has changed status, and the insult it now bears is Silver Wings! The explanation for this sad change is very cruel.

By a rough-count I see that seventy-six people, including the children, lived in Picton during my childhood, and by today the number stands at fifty-five. Some of the men worked on the land, in the farms and small-holdings; and some of them below the land (and water to be precise) in the Point of Ayr Colliery. Some worked on the land *and* in the pit. As far as I can remember there were six outside the above categories, among them Pitar Jones, who worked on the railway; John Roberts, Plas Iolyn the carpenter; Tommy

Williams (Tommy'r Ardd Ddu), a sailor, who was killed at sea by enemy action on 23rd July 1940. In the same grave lies his brother, Leslie Williams, killed in the same enemy action. In the adjoining grave to Tommy's lies 18 year old Ronald Barnard of Ffynnongroyw, also killed in the same action. A sad day indeed for the locality and both families were left 'in a void because of war'.

The veteran Dafydd Griffiths lived in the 'Counting House', the cottage that was attached to the school wall. In his old age he had become rather scrooge-like, because when the balls from the school-yard bounced into his garden they would never be returned; he would keep them in a basket in the back-kitchen. This octogenarian was related to Lena Jones who lived in the farm near our house. Occasionally I went with Lena to visit her grand-dad. When he went out to the garden for some reason, Lena would creep into the back-kitchen and stuff two or three balls inside her bloomers, and then, quite soon, would bid goodbye—mission accomplished. To a lad who enjoyed kicking an old tin at times, Dafydd Griffiths's house resembled Jack Sharp's shop.

The other two, Robat Owens and William Roberts (Wil Wyllt) worked on the land and in Mostyn Ironworks, a journey of three miles each way. Robat would snail-crawl there on his bike, but William walked every step. Robat was not a picture of good health, I believe he suffered from a stomach ailment, and it was rumoured that he enjoyed 'Rennies' more than food. Conversely, William Roberts was stocky and strong, and at times very cross. There was nothing wrong with his stomach except that it had the depths of a pit-shaft sump. He cleared seams of food, because a body like his demanded sustenance for two. While the miners of Point of Ayr were satisfied with a snappin-tin full of butties to eat at work, this J.C.B. devoured a basketful. I saw him a hundred times high-striding past our house carrying his basket with a lid under his arm. I never saw him in Chapel, but his wife Jane, Lasarus Preis's daughter, was a very faithful chapel member in her youth. It is said that Jane was one of the most beautiful girls of the neighbourhood, but later went to live a very withdrawn life. She wore an old sack over her shoulders, and I was immensely frightened of her when I went there to buy eggs, although she was completely harmless, and as frightened as I was if I only knew the truth.

When most Welsh people talk about their locality they refer to their square mile. Mine is a triangular mile. Draw a line from Picton Cemetery to Penyffordd and you have the base of the triangle; the apex of course, is the Ayr Point where the Point of Ayr Colliery stands. Connect these, and you have my locality. I tasted the greatest experiences of my life within this triangle. I was born and bred here, I have spent my life here, and it is here that my body will be laid to rest, in the same grave as our beloved Ennis, in the same graveyard as the rest of the family, David Lloyd, Thomas Edward Jones, and all the choristers of these parts. Yes indeed there will be some magnificent singing with 'the angels way up yonder'.

In the National Eisteddfod the bearers of the Aberthged (sheaf) refer to our national land and earth. Yes, they have my allegiance, but to me there is something far more special in Picton's land and earth. Was it not from this land that I received my healthy sustenance?—Poison-free produce from the garden under the old system of fertilizing, potatoes and turnips as big as footballs from Dafydd Jones and William Parry, and blackberries like jaffas from the Brêc fields; dandelion leaves and nettles for dear Auntie Dol to make the most wonderful herb-beer that was ever bottled. It was this land which gave such a wonderful taste to the beastings we received by the bucketful after one of Dafydd Jones's cows had calved. During these free-flowing days two and three helpings of puddings were commonplace. Because it was I who had taken milady to the bull a few months previously I believe that this was her way of thanking me for the favour! On this land I learned to walk, run, play football, play cowboys, and to love in earnest. Before attending school I learnt one of geography's most important lessons because I knew of the fields by name. It is said that it is difficult to remove a

The Cartrefle family *c.* 1929.

man from his family; yes truly, and it is also as difficult to remove him from his fields. On Cae Bryn the footballing feats were performed. Like many of my contemporaries I did some of my courting in Maes Mari. This place insisted on being included in verses which I composed to two of my friends on their wedding day nearly forty years ago:

> Farewell to lush Maes Mari,
> and Tan-y-Bryn at night.
> The one you helped through courtship
> today is dressed in white . . .

Into this triangle, a part of Wales where the Welsh language flowed as smoothly as the Garth Stream, I was born on 29th August 1926. There were three girls in the family before my arrival, and in two years another son joined the household. I was born during the miners' strike, and my Father like every other union member worth his salt was out of work. Some ups-a-daisy workers went to work under police escort, but they were lacking in numbers; they were also lacking in back-bone. They were never forgiven by their fellow-miners and I remember many a quarrel finishing abruptly in the pit with the words, 'I thank the Almighty God above my head that I never had to come to work between two policemen'. And there was great silence!

My sister, Evelyn, was born in 1921 during a previous strike. No children's allowances during those days. Flint-hard days, and until her grave, when Mam talked about the strikes that had been she referred to them every time as the days of Evelyn's strike, or the days of Einion's strike. I heard about Princess Anne referring to the arrival of children to the family as an occupational hazard. To Mam and my Father their arrival was a blessing at all times—even during a strike.

Tall Hopes in a Little Home

'I'll make it a palace for you, Lal Bach.' Those words were uttered by my Father, Alfred Evans, to his second wife Leah when he brought her and two little girls from their house 'Rhewl House' in Rhewl Fawr to the little one-storey cottage 'Cartrefle' in Picton, half a mile away. He possessed the imagination of a literary man, and a poet's licence, to utter such a hyperbolic statement, bearing in mind the raw materials he had at his disposal—a small white-washed stone cottage, the living-room in the middle and a bedroom on either side. The front door faced towards the East, and it was not even protected by a porch; the back door was conspicuous by its absence. Under the weight of the years the roof sagged in places, but at least it was watertight. Large stone flags were placed side by side to form the floor, and the sun shone through one different shaped window in each room. At one gable-end of the cottage a stone hut had been erected, and adjacent to that was the primitive lavatory, known to us as the petty. The front of the cottage boasted a small garden, one of the gable-ends touched onto a medium sized garden, and the back was dominated by a massive garden which narrowed as it went towards Top-yr-Ardd corner. In one corner of the garden stood the pig-sty, one of the essential commodities of the time, which was far more useful than the modern unnecessary extensions. Putting a pig in the sty was far more important than putting on the style.

Having been born in Rhuddlan my Father was orphaned at a very early age. His father, Robert Evans, was unable to write, and I remember seeing a x where his signature should have appeared on my father's birth certificate. My father's mother's maiden name was Rebecca Bentley. As a result of an accident in the local quarry Robert Evans was killed, and the knock proved to be too severe for the young Rebecca. My father was adopted by a Christian widow, Marged Parry of

Picton, who already had three daughters of her own, Jane, Lil, Dol and Mary, but this did not deter her from giving three other children a Christian upbringing—my father, Magi Francis and Dashwood Cato. After growing up Magi Fransis went to Liverpool, Dash emigrated to America, but my father stayed in this locality all his life. When Little Alfie came to Marged Parry, he spoke English, and was unable to utter one word of Welsh. However, very soon the Welshness of Picton had flowed over him, but not before he had made a howler whilst reciting his verse from the Bible in Gwespyr's Capel Bach. Even Auntie Mary (my father's sedate sister) would smile when she would recall the occasion to us time after time through the years.

My father's first wife was Magi Hughes from y Nant, who worked as a maid in Bryn Mawr Farm, the most prosperous farm in the locality at that time. Bryn Mawr was situated between Glan'rafon and Rhewl Fawr. Although Magi had been brought up in the Wesleyan Chapel, Bryn Seion, Mostyn, when she started work she then 'fell in with John Calvin's cause' in Gwynfa, Rhewl Fawr. My father also attended the same chapel, and after a period of courtship they were married and made their home in a small cottage by the name of Tan-y-Bryn, for a short period before moving to 'Rhewl House'. They had two or three children, who were either stillborn or died within days. Another daughter, Leah, was born, and she lived. However, the mother was not allowed to live long enough to see this little baby grow into a beautiful girl with golden curly hair.

Many years later I saw a record in my father's handwriting in Pitar Williams's Bible referring to his wife's death. It was there also I saw an englyn (a verse in strict metre) in memory of Magi by Owen Owen, Y Brithdir, Dolgellau (Auntie Jane's husband). Lower down the page my father showed his belief in God's Purpose, because where he had noted Magi's death he said, 'but God sent her sister Leah to be my wife, and a mother to little Leah'. I would not be surprised if his belief in God's Purpose and Providence were in this instance pretty near the mark, because he had an excellent wife, and Leah and the rest of the children had a first-class mother. She had also lost her parents when she was a small child, and was brought up by her uncle, Robert Hughes and his wife. Like many of her contemporaries, at the age of fourteen she entered into service in Liverpool. Carrying her hamper, she boarded the train at Mostyn station and was met by her new mistress in Liverpool, where she remained until she came to Rhewl House after Magi's death. It was there that Margaret Evelyn was born, and sometime around 1923 the father, mother and the two little girls moved to Cartrefle, Picton where Rhiannon, Einion and Tudor Wilson were born.

'I'll make it a palace for you, Lal Bach', was the promise, but strikes, and the living conditions of the Point of Ayr miners at that time made it impossible for him to reach that goal. Dear Mam in a palace! She would be unable to issue instructions. Mam of all people in costly clothes! For a while when the children were small she was unable to go to chapel on Sundays because her coat was too frayed, and there were no posh people in chapel those days. Oft-times have I heard her say that she made sure that Dad had his best suit for Sunday and that all the children were tidy to go to worship. Mam had to wait until she had an old coat from Auntie Lil or Auntie Dol before she could re-start going to chapel. After that she did not miss going to chapel all her life. My sister Evelyn's death from T.B. at the age of twenty-four was a heavy blow to the family. Evelyn was the dearest of us all. She was the only one in the world who could make my father change his mind. During the bereavement my father stopped going to chapel for months, but Mam went there regularly twice each Sunday. Hers was a quiet, deep, endless faith, and through love and tenderness she was able to persuade my father to return to worship on Sundays—'and both re-entered Bethel'.

The first step towards turning the cottage into a palace was to erect a porch on the front door as a little extra shelter from the winter storms. Llannerch-y-Môr supplied the timber; I believe that there had been a fire at Robert Davy's Yard and a lot of timber was given away free to those who would supply the cartage. My father never failed to take advantage of a bargain, and he soon borrowed a horse and cart and brought a massive load of timber home. He then set about erecting the porch. Neither did he pay for the nails. Those came by large fist-fulls from some underground warehouse whose name I cannot recall at the moment! It is also more than likely that the hammer came from the same warehouse. A door was fitted to the porch, and two small windows, one facing the road, and the other facing the path that led down the Brêc. After completing the job, which was, by the way, quite decent, it had to be painted with a dark green paint, which was to be its colour practically all its days.

After a few years a back door was essential. At first, it could be rather awkward, because one had to travel through the front door even to get into the back garden; and sometimes it could be a bit of a crisis when an emergency call was sounded and a quick access was needed to a rather special little place. My Father was the architect, joiner, stone-mason, and labourer.

'Cuppa, Lal.'

Whatever Mam happened to be doing at the time it would be 'down tools' immediately, put the kettle on the fire and prepare the cuppa which included, more often than not, a couple of slices of rhubarb tart, thick slices of *bara-brith* and two cups of tea with lots of sugar. At times like these Mam would be lavish with her sweeteners—

'It's looking lovely, Dad.'

Then another stint until dinner time. After levelling the food, he would have some seven hours sleep before going to work on the night shift at the Point of Ayr. The night shift was his lot for years and years. When he asked Mr Young, the Manager, for a change of shift, the praiseworthy but useless answer he got was, 'Oh no, Alf, you're a nightingale.' Returning about 6.30 in the morning, to a fried breakfast, a top-half wash, and then re-start on the contract. Not only was a back

Mam at the cottage.

door provided but also a back-kitchen, one with dry walls (the wages were too small to afford mortar), and because there were five children running about the place it was unnecessary to have a specially made floor. The earthen floor would be as hard as a pyramid's floor before one could say 'Tunnel Cement'. With a small window, and a galvanised sheet roof, the extension was at least useful.

'This Old Hut at the gable end is too small for us now, Lal.'

'Yes, I'm sure, but where will we get the stones to build another hut, Dad?'

'D'ye see this wall in front of the house? I'll pull it down and replace it with timber from the Brêc, and use the stones to build the hut. Mary, my sister, has had a new back door. I saw the old one in the garden when I went to mend Marged Jones's clock the other day; and Will Parry, Tan-lan, has had a new hay-shed, after pulling the old one down—like that man in the Bible, Lal. I'm sure I can smooth the old Will's hair to have half a dozen sheets from him.'

'Well done, Alfred the Great.'

That was the sweetener for my Father that day. That, simply, was how the 'New Hut' became part of Cartrefle. These improvements entailed a lot of hard work for my father, on top of his work in the pit—cultivating every inch of the garden, keeping a pig, keeping hens, going around pegging, and killing pigs, repairing our boots, helping in the temporal harvests, and attending the spiritual harvests five times a week.

Apart from the eventual addition of an unsuccessful Elsan in the petty, and a few other minor alterations, I believe the palace-resemblance exercise came to a halt somewhere at this point.

How I Saw the Picton Scene

'Dinner mam-mam, dinner mam-mam,' and my cousin Megan knocking our dinner table with a spoon. She had learnt to say 'Mam', but in those days she could not say Auntie Leah, therefore to Megan, Mam became mam-mam.

'Dinner, mam-mam.'

Knock, knock, knock. Megan was not the only one to knock the big, solid table! Toys were always very scarce in Cartrefle. Strangely, a few weeks before each Christmas, a terrible catastrophe would take place somewhere, according to Mam, and it was not possible for Father Christmas to bring us many presents. We accepted the situation, we had been taught from the early days that the children who lived 'Far, far in China, and Isles of Japan' were much poorer than we were. Had we not seen Mam putting pennies in the little box on the dresser? We would also go around Picton each year with our pink cards to collect money for the little far-off children. By today much is made of the idea of twinning towns in Wales with towns in other countries. Without realising it, Mam succeeded in twinning Top-y-Brêc and the Khasia Hills in India, and that without any help from any multi-medallioned Mayor. In our view all would be well if those children received most attention from Father Christmas. Our proximity to Dafydd Jones's farm was the reason why we had no fireworks on Bonfire Night: there was a danger that the hay-stacks would be set ablaze. But if toys were scarce there was an abundance of nails in a box in the corner where my father's pit-clothes were kept; there was also a hammer there, and where better for a toddler to knock them than into the big table? I hammered dozens in, most likely when the room was empty. It's not possible that my loving parents would approve of an action like this, but neither can I recall being scolded for doing it.

Knock . . . knock . . . knock . . . knock . . .

Hammer . . . nails . . . signs of the cruel capitalist system and the suffering of the twenties.

Less than fifty yards from our house, in Dafydd Jones's stack-yard there were two small circular hollows in the ground, the remains of the two shafts of Picton Colliery. The shafts had been made safe and Tiw and I played football there a thousand times. They must have been safe, because Mam never warned us about playing there. She continually warned us about going to play by the red water on the Brêc, because of the great danger of being drawn into the mud, but many is the time she watched us playing and lying on the remains of the old shafts. Last year, a large hole appeared in one of the shafts for a depth of fifty yards. Fortunately no one was playing there at the time. Opposite the shafts there are three houses, Glasfryn, Cartref, and Hafod. When coal was wound up from this pit Glasfryn was the office, and to this day a stone staircase rises from the living-room to the bed-rooms. Auntie Dol, Uncle Twm, and their children, Sam, Parry, Joe and Megan lived in Glasfryn, a widow named Marged Jones lived in Cartref, and Auntie Mary, also widowed, lived in Hafod. These three houses together with our house formed part of the old coal yard, and the place was referred to as Picton Yard, and we were quite happy to be called 'yard people'. The yard led to the Brêc footpath, which in turn led to the Point of Ayr.

'Close your fist tight, and hurry home.' Those were my cousin Joe Blythin's words to me when I went there as a small boy on a Friday afternoon to fetch my father's wages. The Point of Ayr miners were paid at the end of the morning shift on Friday. For those on the night shift this meant a special journey to the pit, disturbing their sleep, or having someone else bring the wages home. Joe Blythin was my father's Sec-uricor van for years, and very often it was I who brought our part of the load for the last fifty yards of the journey. Two pounds ten shillings was the amount I clutched in my fist, and Mam would perform wonders with it when she got it into her purse. Regardless of the state of her Dow Jones Index it was essential for her to get an ounce of shag tobacco for my father to smoke, and an ounce of twist tobacco for him to chew in work, and an occasional little chew whilst digging the garden. This was his weekly ration for years. He would increase his incense to Bacchus by drying mounds of colt-foot leaves in front of the fire, rubbing them slowly between the palms of his hands, and mixing them with the tobacco. A few pence a week went to the insurance man in order to have enough money for funeral expenses, and the remainder went to pay for the house, food, and clothing. Our food was plain, plenty of garden and farm produce, sheep's-head broth and home-cured bacon. Quite often old Lei Burum would call by with half a dozen rabbits swinging on the barrel of his gun, and on that day Mam would make us a mouth-watering dinner for about sixpence. Through the week we would have white bread, but without fail *bara-brith* would be placed on the table every Sunday. In order to push the *bara-brith* further Mam and the children would have a *pioden*, that is a piece of *bara-brith* and a piece of white bread together. Cartrefle's Head of House did not believe in a *pioden*, and he would have two pieces of *bara-brith* together every time. To sweeten a little on his first shift every week he would take *bara-brith* for his snappin on a Sunday night. At the time I felt annoyed because he had more of the *bara-brith* than I had. Sorry Dad.

We would have Christmas cake if Will-y-Bara, Gwespyr gave us one as a present. Thank you, Will, for putting icing on our festival. We never saw a turkey on our dinner-table but we had a brace of pheasants one Christmas. At one time Auntie Mary nursed an old gentleman called Mr Graves in Bron Eifion, Criccieth. Before this particular Christmas Mrs Lloyd y Post delivered a box to our house. On the exterior of the box printed in large letters was the word 'GAME'; 'Hooray, Auntie Mary has remembered about us and has sent us a box of games.'

'I'll have the snakes and ladders.'

'I'll have the Ludo.'

'I'll have the draughts.'

'And we'll all have the pheasants,' said my father laughing hilariously. Because 'Penrhiw's donkey was on my back' I did not have an appetite for the pheasants when they were put on the table for Christmas Dinner, but Will-y-Bara's cake tasted exceptionally good when tea-time came around.

Because the four men in Auntie Dol's place were miners they had a few more relishes than we did. On Sundays they would have tinned fruit and cream for tea, and during the week they used quite a lot of condensed, and evaporated milk.

I remember going down often to their rubbish bin to gaze at the colourful, mouth-watering empty tins of peaches and pears. Desiring them, but not tasting them. Another of the luxuries in Auntie Dol's house was Oxo. I remember this with a clarity associated with Plas Robin well. At that time each oxo was individually wrapped in a small cardboard packet, and if someone collected a hundred of these little packets, they could be exchanged for a doll or a football. Fair play to the best Auntie who ever wore a shawl, she collected them without fail and sent them away, for the mischievous, lively lad from the top of the yard to receive a spanking new football from Mrs Lloyd y Post, of Ffynnongroyw who delivered the mail for years in the locality. A football for Islwyn and myself to be real footballers. Although born in Picton Hall Farm, Islwyn had moved to Gronant at a very early age, but he came back to his grandfather, William Parry's farm very often, and we played a lot with each other. Usually, at best, we only had a small ball, and sometimes when that got lost an old tin from Auntie Dol's rubbish bin would prove quite acceptable. Embracing and kicking a size five football at times was a wonderful experience. It wasn't once I received this precious gift, but many times. Auntie Dol will be special in my eyes for ever, and I know that I was also a little special in her eyes too. To her, I was a blue-eyed boy with curly hair, and if anyone wronged me she would be the first to 'put salt in his broth'.

One day I swaggered down Rhewl Fawr with the football under my arm. I was as proud as any captain who ever led his team out at Wembley. I met a big lad there. His father also worked in the pit, but received a bigger wage than my Father.

'Y'uve had a new football.'

'Ooh, yes, from Auntie Dol.'

'That's an Oxo football, I've got a much better one. I've had it with Bovril coupons. Because in our house we eat Bovril, not Oxo.'

I did not bother to tell him that we did not even eat Oxo in our house, only onion and dripping stock.

'Carry on, Dixie Dean'.

In order to raise his family's living standard to the same level as the Oxo eaters, and maybe, some far-off day, to the standard of the Bovril eaters, my father decided to open a shop in the cottage. Yes, honest-gospel. Seven in the family, only two bedrooms, and one of those was turned into a shop. The counter was a large table, and because the walls were thick there was quite a wide shelf on the inside of the window. There, out of our reach the sweets were kept—the bon-bons, the Uncle Sam's Toffees, the mint-imperials ('the prayer-meeting sweets, Einion' according to Twm Llai Mên, Point of Ayr's champion swearer), the caramels, and the liquorice. Room was made on the table for jam, butter, salt, pepper, soda, donkey stones, washing blue, and other small requisites. Near the table the vinegar barrel was kept. 'This vinegar is very strong, Mrs Evans, and you can dilute it quite liberally with water before you sell it to your customers,' was the traveller's advice.

'Oh no! I could never do that.'

'Do you know what your trouble is, Mrs Evans? You're too dammed honest for this world.'

The venture was named Cartrefle Stores. Sparse was the stock. Sparse the customers. Sparser still the profit.

'Alfie.'

'Yes, Mary,' said my father to his sister.

'I think it would be a good idea for you and Leah to start selling paraffin. You know that we all have to walk at present to Brooks' Shop in Penyffordd to obtain paraffin, and a gallon is a little heavy for those who are widowed. When one reaches Hogla Teim corner the arms begin to ache, and by Cae Olyn the can must be put down for a while. Having the paraffin here would be a great help for all of us.'

'And where am I to store it, in the pot?'

'No, I thought you could get a fifteen gallon tank from the ironmongers in Prestatyn. I think you could get one there without being too dear.'

'Mary bach, I was at my best buying a three pint tin-bottle from Hugh Tincar a fortnight ago, to carry tea to work. I had to do without chewing baccy for a week to have enough money to pay. After that I was cadging baccy all week. Buy a fifteen gallon tank? Out of the question, Mary, bach.'

'Listen, Alfie, you know that I receive a pension after dear Noel was killed in the War, and that I have a little money put by. Do you remember Mam, Alfie, teaching us to save for a rainy day. Oh, dear Mam. Listen, I'll buy the tank and you and Leah can repay me at the rate of half-a-crown a week.'

For the sake of peace, and not because of any business acumen which my father possessed, a branch of the Cartrefle Stores was opened in the Old Hut, and a large green tank with a small pump at the top was installed there. Double farewell to the fireworks after that. The 'branch' lasted appreciably longer than the shop, so for a time, it was my Father's responsibility to supply the people of Picton with light in the pre-Manweb era.

> 'The children, though poor, are healthy and tall
> because she worked wonders on wages so small.'

I learnt that couplet about the miner's wife many years ago. Of course, it was true about most of them. Wonders of feeding, wonders of clothing, wonders of hiding the worry from the children. I believe Mam had an added wonder, finding room for all of us to sleep. Because one of the bedrooms housed the 'Stores', everyone slept in the other bedroom. Both beds and the cot must have been in the same bedroom. Because my Father worked nights things were not too bad until Saturday night, when he did not work, then we could truthfully say, 'O master seven are we' as we crammed into one bedroom. I cannot recall the Cartrefle Stores ever selling sardines, suffice it was to rear them. As we children grew a little, the situation was impossible, and I do not think it caused my parents any regret to turn the Stores into its original function. Mainly as a favour to the neighbours the 'Branch' was continued in the Old Hut.

Pre-Manweb days, pre-Elfed Jones days as well. Water had to be carried from Plas Robin well about a quarter-mile distance away. Eveyone with his buckets and wooden frame. Everyone except Dafydd Griffiths, Counting House. He lived about three-quarters of a mile away from the well, so he put two wheels under a small barrel, slapped a handle on the barrel and pulled it backwards and forwards to the well. 'Cometh the hour, cometh the man'.

During this period Major Bates came to live in Castell-y-Gyrn, about two miles from Picton. He bought some of the Picton farms from Sir Piers Mostyn of Talacre, although I'm sure the change to the tenants would be negligible. The rent-dinner remained,—as did the poverty. Nevertheless I'm obliged to put one feather in the Major's cap; if liquors flowed freely in Castell-y-Gyrn, he made sure that his tenants had water. He dug a small water reservoir in Picton, and water was piped to the surrounding farms. Because of this we also had a water tap near the gable end of our cottage. After this it was farewell to Pwmp Plas Robin and Dafydd Griff's barrel.

The stop-cock for the tap was placed in our back garden, near the gooseberry bush, about four foot underground. It would be reached by raising a small trap-door, and by using a six foot iron bar with a spanner at one end the stop-cock could be opened and closed. The safe-keeping of the iron bar was the responsibility of the proprietor of the Cartrefle Stores, and it was his (unpaid) duty to switch the stop-cock off when there were signs of a hard frost. On these Arctic occasions I was sent down the yard and to the farm to tell everyone to fetch enough water for the following day. In about half an hour my father would carry the big iron bar with as much dignity as Gwyn Dob carried the Gorsedd Sword during the National Eisteddfod. During adverse weather this was a daily ceremony.

Mam looking after the vinegar. My father looking after the water. And both of them were as adamant as each other that they should never be mixed in the Cartrefle Stores.

A Small Snap of Both Chapels

Picton has no chapel, but Gwynfa chapel stands on the outskirts of Rhewl Fawr. A part of the Calvinistic cause, a visible part of the 1905 Spiritual Revival. Downwards, at about another quarter of a mile towards Penyffordd stands Peniel chapel, where the Wesleyans have worshipped since 1899. Because of their religious success they built a much larger chapel in 1926, and turned the little chapel into a schoolroom which became quite useful in the locality.

In his childhood my father went to Capel Bach, Gwespyr, but in 1905 he came to Gwynfa chapel. This meant that he had to walk considerably less to the services and frequent meetings. I cannot recall ever seeing Dafydd Jones, Picton Yard Farm in chapel, although he treasured all the hymns in his memory, and his family were regular worshippers, but I've often heard him proudly boasting that he carried massive

Gwynfa Chapel (Calvinistic Methodist) Rhewl Fawr.

loads of stones from Gwespyr Quarry with a horse and cart in order to build the chapel. The stones were brought from Gwespyr, the 'living-stones' had previously been shaped in regular meetings in Rhewl Fawr Farm's parlour, when Charles Williams lived there.

Before Gwynfa Chapel was officially opened in September an important evening was held—an opportunity to choose and pay for a seat. I remember Tommy Morris, my father-in-law, relating to me an incident about his father John Morris going to choose a seat for the family. John Morris lost his wife when he was a young man; he himself sustained a terrible accident in the Point of Ayr where his back was broken. He was crippled for the rest of his life, but he reared the six children himself. There was no compensation those days, except for an occasional bucketful of butter-milk he received from a farm associated with the pit. The owners bought him a wheel-chair,—big deal—and threatened to take it from him during one election unless he voted for the Tory. John Morris kept his chair, but the family knows full well on which part of the paper his cross was placed.

It was in his wheel-chair that he came to chapel to choose a seat for the family, and to pay the Seat-Collection for the first year. Travelling to chapel was a great effort for him. Finding the money to pay was also a great effort. My wife Betty's family has occupied this seat without interruption from that time, and Betty and I still sit in it to this day. By now only a couple of families in the chapel bother to pay for their seat each year—it is only ten shillings per year. Through our lives Betty and I will pay for this seat, mainly as a tribute to her grand-father's sacrifice.

You may have heard it claimed that Emlyn Williams the actor was the first baby to be christened in Gwynfa. If the baptismal register is correct (and I have no room to doubt its authenticity) then he was the second. The first name on the register is Amelia Williams daughter of William and Elizabeth Anne Williams, Rhewl Fawr, and the christening took place on 12th December 1905. The officiating

No. 1 Jones' Terrace. Birthplace of Emlyn Williams

clergyman was the Reverend O.B. Jones. The 'elder' of the stage cared very little about his fleeting connection with the Big Seat in Gwynfa Chapel—at least he showed no interest whatsoever when the half centenary of the Cause was celebrated, although he had been informed of this important occasion by the secretary (or maybe by the treasurer!) During the chapel's early years they did not possess a proper christening bowl, but during christenings a small bowl was borrowed from Mrs Frederick Parry, Rhewl Villa. In later years Mrs Parry's daughters Lilla and Dilys presented the bowl to Emlyn Williams.

Because I have ventured to correct one fact about Emlyn Williams I might as well carry on to make a correct note of his birthplace. I have seen references to him as one 'born in Holywell', 'born in Rhewl Mostyn', or 'born in Trelogan'. The simple, correct fact is that he was born in No. 1 Jones' Terrace. The author of *The Corn is Green* was born there in 1905, when the Sheaves of Grace were being carried in golden loads to Gwynfa Chapel. To complicate the work of future researchers this information can be seen above the door of No. 6:

Jones' Terrace 1907
E.J.

There are eight houses in Edward Jones's Houses (as referred to by the locals). The five lower houses were built at the turn of the century, and the three upper houses were built in 1907, thus giving rather misleading information above the door of No. 6.

In Gwynfa three ministers were raised from the seed of the Sheaves of Grace, namely the Reverend Hywel Jones, Gellifor, the Reverend Ifor Oswy Davies, Liverpool, and the Reverend John Herbert Roberts, London (later of Abermiwl). These showed quite an interest in the celebrations marking the half-centenary of the Cause, the latter two preaching powerfully on the occasion.

'Ev'ry man is bound to show
the place where once his roots did grow'

Gwynfa was never a great chapel in size nor attitude. The greatness, in every respect, belonged to the 'mother-church' Moriah, Ffynnongroyw, in size similar to a temple, and the massive pipe-organ resembled the Spion Kop in Anfield. In the Preaching Meeting which was held on the Thursday night and all through Good Friday, two ministers officiated (Calvinistic Methodists of course!) and on those two days, on the table below the pulpit, the large white lilies would shine like Talacre Lighthouse. From Moriah also two brothers (in the religious sense) would occasionally come to assist in our prayer meetings after receiving a call from our Macedonia. I have reverred memories of Thomas Edwards, his brother Isaac, and James Parry coming to us, with the dews of the revival still refreshing their souls. By now, more is the pity, the greatness of Moriah has long gone, leaving a legacy of problems. The visitation of the Holy Spirit has now given way to the visitation of unholy vandals.

In the obituary reports in the bygone *Prestatyn Weekly* and *The County Herald*, everyone, except the pub landlords were referred to as 'a faithful member' of some chapel or other. My father was highly amused to see this description being given to people who had not attended the chapel since the Revival. However, we as a family, were in truth, faithful members of Gwynfa Chapel. With his rich, deep bass voice it was natural that my father led the singing. We attended three times on a Sunday, went to the Band of Hope on Monday, and to the Society (or prayer meeting) on Tuesday night, my father having to change immediately to his pit clothes on arriving home on the three nights.

In those days I went quite ungrudgingly to chapel, though the Society appealed less to me. The singing was immensely enjoyable, but the reciting of my verse from scripture had the taste of wormwood-tea. It was a tradition for the boys to stand at one side of the 'Big Seat', and the girls on the opposite side. After tea on Sunday I would occasionally commit a new verse to memory in readiness for the evening service. Because the verse had not been given 'depth of soil' in my memory, it would disappear when I heard the Minister say 'and the next'. By some strange miracle salvation was close at hand. Opposite to where I stood a commemorative plaque had been placed in Memory of Mr Thomas Pownall, one of the founders of the Cause, and at the base of the plaque were the words: 'Enter thou into the joy of the Lord'. This would be 'sufficient help for me in my adversity'.

It was easy, afterwards to remember the whole verse. 'Well done, thou good and faithful servant, thou hast been faithful over a few things. I will make thee ruler over many things; enter thou into the joy of thy Lord'.

I was fortunate to be standing on the boy's side, or my contribution would very often have been in the form of a blank verse.

In my youth I continued to attend chapel; the habit of turning your back on the chapel as soon as you had been confirmed had not started at that time. There was also an added attraction for me at Gwynfa, because Betty, and her family were also 'faithful members'. More often than not I reached Cae Bach stile at the same time as Betty, and we would walk arm-in-arm to chapel. It matters not that every word of the sermon was not enjoyable, the words in the 'friendship meeting' later (*Y Gyfeillach* or *Seiat)* would be beyond compare.

In the Band of Hope Mr and Mrs Williams the Minister would teach us short plays, my father would conduct the cantatas, and Mam would swell ever so little on the choruses in the practices. The concert would be held at the end of the Winter, on a make-shift stage in the 'Big Seat', with borrowed curtains from Peniel Chapel, and at the rear of the stage a beautiful, coloured banner belonging to the Rechabites (a Temperance Movement). In those days all chapels supported the cause of temperance; free-souls were far more important to them than free-masons.

I recall the joy of learning the cantata, *Hope of The World*, a rather special work, because a hamperful of the national costumes of the different countries was borrowed for the concert. Before the big occasion the hamper arrived at Mostyn Station, about three miles from our house, and my father went there and carried the load home on his shoulders. After the performance the hamper of costumes was transported on the same shoulders to Trelogan for the little children of Disgwylfa Chapel to sing about the little children who lived 'Far, far in China'. The hamper was again conveyed on the same shoulders from Trelogan to Mostyn Station. My father revelled in his strength and bravery, little wonder that at times Mam would call him Carnera.

Not so pleasant my recollection of an incident one night when I was about five years of age in the Band of Hope learning *The Bird's Cantata* with its

> Now hear the wren's song
> so loud and so long.

In the song, mention is made of 'the little wren's flute', and my father asked, 'Where is the little wren's flute?'

Bob and Dei, the twins, sat behind me. They were full of mischief all their lives. Surnamed Edwards, they were related from afar to Twm o'r Nant. (The man from Nantglyn who wrote and acted in interludes in his day). One of Twm's family who was engaged in carrying timber from Prion to Mostyn Docks came to work as an ostler in the Point of Ayr. Fair play to Bob or Dei (it was impossible to differentiate between them) for whispering the answer in my ear.

'Where is the little wren's flute?'

'In its bum.'

Glory be, my father went wild, and he forced me to go to sit in the pulpit, far away from everybody. I wasn't fit to be near other children. The rest of the evening was spent crying in the pulpit, until my eyes were as red as those of a herring.

Mr Williams the Minister left the Band of Hope that night immediately after the explosion. Because he was overcome with shame according to my father. And under the weight of my shame I had to go to bed immediately, without a bite of food until morning. At five-past-nine, after my father had gone to work, Mam came to the bedroom with a glass of warm milk. A few years later I learned why Mr Williams left the Band of Hope so suddenly—he couldn't control his laughter.

Years later in 1964 I won the Anglesey Eisteddfod Chair in Aberffraw (not the National Chair). By this time Mr and Mrs Williams were shepherding the flocks in Gwalchmai,

Anglesey. Immediately after the ceremony Mrs Williams came to me, as proud as punch, smiling from ear to ear, and her first question to me was:

'Where's the little wren's flute, Einion?'

Then she introduced me to her friends as one of the Band of Hope boys from Gwynfa.

The members of Gwynfa and Peniel believed in joint-services on some occasions. In the past, for Harvest Festival, and for the services during the first week of each New Year, but by now, because of small numbers, through August. I never saw much difference between members of either chapel. My mate, Don Brooks, attended Gwynfa, and my other mate, Alun Myrtle House went to Peniel. I was perfectly at home in both places.

One small difference which I noticed between the denominations was the unchangeable pattern followed by the Wesleyans in Prayer Meetings. At this time it was the men who took the devotional parts, the women did not pray publicly (although being mostly miners' wives they would have prayed often in fear, and in secret). I have heard it said that my grand-mother, Marged Parry, prior to this time, prayed in public, with a fluency and a spirit which left the men in amazement.

This is the procedure which I noticed at all times with the Wesleyans. When the second member would come forward to pray he would always say:

'Our Father, which art in Heaven, behold us now approaching the Throne of Grace in prayer, after our dear brother.'

When the third came forward to pray he would come 'after our dear brethren in prayer'. The fourth one that came forward also came through the same gap in the hedge. I know that they quarrelled often in work, but before the Throne of Grace they were all dear brethren.

For years the W.E.A. classes were held in Peniel vestry. Usually the local clergy acted as tutors, and I know that they could all do with the little extra money. I have absolutely no complaint whatsoever regarding the tutors, they laboured hard in the locality and they fully deserved a little bonus from somewhere, but I doubt the wisdom of their choice of subject at times. One of the topics was 'The Philosophy of the Greeks' and to follow this unpalatable course, 'Psychology' was put on the educational menu. I cannot imagine my Father, Bob y Garth, Huws Gabriel, William John Portar and the other dear working-class members becoming very great chums with Aristotle, Plato and Freud. A snappin-tin full of the poetic rudiments would have been far more appetising. William Courtney, Herber Jones, Alun Jones, Abel Jones and Gwilym Parry possessed the bardic seed, it is a tragedy that no Apollon came to the locality in their day.

It was in Gwynfa Chapel that Ennis, Betty and myself were christened, and confirmed as members, and it was there that Betty and I were married. For years the worship was enriched when Ennis and Betty officiated at the organ with dexterity and feeling.

It is said that the time is coming when the Father 'will not be worshipped in this mountain nor in Jerusalem', but in the meantime Betty and I intend to worship with the 'little flock' on Rhewl Fawr hill.

In Touch with Yesterday's Team

Football has always played an important part in my life, and even now I often dream that I am on the football field. The first recollection I have of playing was one lunchtime as a member of Picton School on Y Weirglodd, a field belonging to William Parry, Picton Hall Farm. Elwyn Morris was the teacher on duty, and during the game I injured my left knee. Mr Morris carried me home, and Mam emptied the bottle of iodine on my injury. By the following day I had recovered fully, and could kick the ball as well as ever.

Penyffordd boasted of a football team in the 1920's, but my only personal recollection of this is the name 'Penyffordd Jolly Boys'. They played on Cae Mawr, a field owned by Hugh Jones, Caeau Uchaf, and I believe that there are many exciting incidents of the period which could be related. I also recall a team from the village called 'Penyffordd Eleven Gents'. This team was never in a league, but played in some knock-out competitions which were held frequently on one of the Point of Ayr fields. A whole day of footballing. Boredom was unknown those days.

I must have been about four years old when I saw Jack Parry, Dee View pass our house post-haste. He told me that Penyffordd were playing against Ffynnongroyw in Y Weirglodd. I followed him at once. Oh! glorious days, when it was perfectly safe for a young child to follow a man whom he only knew slightly.

Four poles from Flanders Woods sufficed for goals. No cross-bars, no white line on the field, but the match was great. I believe that either Jack Parry or Will Dale acted as referee. The result has not stayed in my memory, but I remember Emlyn Preis, while trying an overhead kick, striking the ball on his own face until his nose was bleeding like a slaughtered pig. To my knowledge that feat has never been repeated.

A while later I recall going to Pen-y-Maes field to watch a team from Penyffordd (not yet affiliated to any league) play against the Holywell County School team. By now crossbars had been added but not nets. The only place to change was in the revealing shelter of the hedge. The County School boys travelled the seven mile journey on their bikes, accompanied by the monocled Mr Christopher. Three local boys, Dewi Morris, Iorwerth Jones (Pen-y-Maes) and Albert Henry Parry played for the County School, who lost by about six goals to nil. Soon afterwards Penyffordd joined the Dyserth League and things became more organised. Nets on the goals, markings on the pitch, and changing facilities in a room in the White Lion Inn, a nearby pub. Footballs were rather scarce those days. Sometimes only one ball was available, and I remember a few games coming to a premature end because the ball had been punctured by being kicked to the hedge. Very strangely, this happened when our team was losing. At the time I could not understand why this was so.

The field was situated above Glan'rafon, being very centrally placed for supporters from Picton, Penyffordd, Llanasa, Trelogan, Gwespyr and Ffynnongroyw. Hundreds flocked there to watch (some to fight at times!) The linesman's duty was very important. He did not wave a flag, only a handkerchief (sometimes of dubious whiteness!) which he waved when the ball was out of play, and he also kept the crowd away from the actual playing area. He paced the line in both directions like a scalded Roger Bannister, crying out 'toe the line, please, toe the line'. Because the linesmen were chosen from the supporters they would be as good as an extra player. Substitutes were unheard of in those days, and if one of the lads was injured the team would have to complete the game with ten men. I often joined in the affirmative battle-cry, 'Come on the ten men'. Another important member on the field was the First-Aid man, with his magic sponge. This post was filled with dignity for years by Dic Williams. When Dic considered some matches to be of some importance, he would attend in his St John's Ambulance uniform. I know that it's too late, but thanks, Dic. The first jerseys during this period were of black and white stripes, then black and white quarters, then red ones. I still hear the crowd shouting, 'Come on, the Robins'. At first, payment was not made by the gate, but in a collection-box which was carried around the field at half-time by George P. Jones who was a wonderful secretary to the club for untold years. Mam would give me a penny, but when the collection time arrived I would scamper over the nearest hedge, where I would squat until the game restarted, then I could spend the penny at the end of the game in John Jones, Africa's shop in Glan'rafon. My cousins Parry and Joe Blythin were some years my senior, and it was with them I went to Pen-y-Maes field at first. It was also to their house that I went to listen to the Cup Final from Wembley every year.

Dear Auntie Dol made sure that one of the lads took the wet battery to Adams' Garage, Tan-Lan, to have an injection of yesterday's 'steroids' to boost its inner mechanism in case the wireless should 'become subdued, and silent be' on such an important occasion. We did not possess a wireless in our house—only five loudspeakers! Arsenal, Everton, Newcastle, Portsmouth, Preston North End, and Wolves were the important teams, in the days when the ball was played forward, and some of the idols were Dixie Dean, Bastin, Carter, Ted Drake, Ted Sagar and Bryn Jones. The only sight of these players I had was on cigarette cards which I received from Parry and Joe and their mates. I gazed at these cards for hours. My cousins took the Penyffordd Football Teams so seriously that when they lost they went off their food. This is perfectly true: Phyllis, Joe's widow, still lives in Picton and can confirm this emphatically. I supported the team staunchly for years, that is when they played at home. A bus, sometimes two, conveyed the team and supporters to away matches; the adults paying a fare of sixpence and the children threepence (the old money). Mam always made sure that she had three-pence for me and the rest of the family for collection on the Sunday, but she could not afford three-pence on a

Saturday. As always, Mam had her priorities in exactly the correct order.

By the days of the red jerseys the games were played in Picton on Cae'r Trumiau, owned by Pitar Jones; the teams changing in one of his sheds, and enjoying the luxury of a wash in a little galvanised bath at the end of the game. It was not customary for the referee to have a wash, he would always be in a great hurry to get away before he was scrubbed! Many memorable games, and some excellent teams were seen in the locality . . . Penyffordd, Mostyn Y.M. Mostyn T.P.'s, Newmarket, Courtalds, Flint Athletic, Holywell St James, Denbigh Mental Hospital, Denbigh Red Dragons, and Llandulas amongst them. Those days games lasted for a fortnight, a week prior to, and a week after the game. Conveying the players to home matches was not very well organised. I recall Alf Smith, Meliden cycling the seven mile journey to Penyffordd, and I believe that, to start with, Eirwyn and

The Old Team *c.*1938.
George P. Jones, holding the Talacre Cup. Hugh Jones, Rogers, Emrys Jones, Tommy Tyson, Charlie Gaffney, David Beatty Edwards, Gwilym Jones (Captain), Bob Edwards, Les Parry, Eirwyn Jones, Reuben Parry, and George Davies the Lodge presenting the medals.

Emrys travelled on their bikes to thrill us. Plenty of memorable games—but one unforgettable game—a Final in Rhyl for the Talacre Cup against Flint Athletic about 1938. The Flint team were hot favourites, because they had practically trounced everyone through the season, but the Robins showed their metal and won by two goals to one, Les Parry Cyrnol and Emrys y Blac scoring the goals. The game is talked about to this day. I missed that wonderful game, the same old problem with that three-pence again, and I was about twelve years old at that time. There have been many good teams in Penyfffordd over the years, but none to compare with the 'Old Team'.

A few years ago I went to Wembley to realise one of my childhood dreams, and saw Liverpool and Wimbledon playing. Believe me, I would have gladly sold my ticket to anyone for three-pence had there been a magician around who could wisk me back to 1938, to the final in Rhyl. Seeing the Robins again, seeing the ball stuck in the Flint Athletic net, seeing Gwil y Farm (the captain) being carried shoulder-high by his team-mates and the Talacre Cup in his hands; being a part of the rapturous harmony of the victorious throng . . . Wembley indeed . . . 'Come on, the Robins'.

A Landmark Stark and So Deep

At one time it could be said that 'All roads lead to Rome'. All the roads and footpaths in this locality lead to the Point of Ayr Colliery. The hills of Llinegr, Tan-lan and Ciwbil were the contribution of the Macadam brothers to the geography of the area, and the official paths of the ordinary folk were Y Rhiw, Y Glasdir, Y Brêc, Gwespyr Rhiw, and the Cob in Ffynnongroyw. No need for the Job Creation lads to tidy these footpaths during my childhood. My father travelled up and down the Brêc to work over the years, until he had a second-hand Austin a short while before he retired, aged sixty six. My father was painfully punctual in all he did through his life. At ten minutes to nine every night he was ready to go to work, his snappin-tin and his bottle of tea in his big poacher's pocket inside his jacket, the watch having been wound. He would have a short smoke until Big Ben started to chime on the wireless (after we had been given a second-hand one from Auntie Dol after 1940). On the first stroke he would rise and give Mam and the girls a kiss, and a small playful chafing with his whiskers to Rhiannon on her face. In those days my father only shaved on a Saturday night.

'Take care, Father.'

'Don't worry, my old Lal, I'm bound to be alright.'

This was the farewell bidding for years. During the war when the German bombers released their devilish loads on Liverpool, and occasionally some of their left-overs in our area, and when an occasional Spitfire snarled at their tails, the farewell bidding changed slightly.

'Take care, Father.'

'God be with you till we meet again.'

Coal has been mined from the pit for about a hundred years, and has afforded a living (truly, a very slender one at times) to hundreds in Ffynnongroyw, Tan-lan, Gwespyr, Llanasa, Trelogan, Newmarket, Acstyn, Picton, Rhewl

Point of Ayr Colliery from the Top of the Brêc.

Fawr, Penyffordd, Mostyn and Trefor. Then, with the advent of the buses, the catchment area was enlarged to include Rhyl, Prestatyn, Meliden, Holywell and Flint and, for one poor pilgrim, Northop. Will Jos, hard of hearing, heavy of clogs, travelled from Northop to Flint on one bus, changed there to another bus to be conveyed to work by 5.30 in the morning, a journey of thirteen miles in one direction. As the man in the Bingo says 'unlucky for some'.

At one time there were many coal-mines in the area—Picton, Plas Robin, Mostyn Collieries, Englefield, Bettisfield, and Wrexham pits of course, as numerous as rabbit warrens. By now Point of Ayr is the only one that is producing coal. How much longer I wonder? There is enough reserve coal to last another hundred years, but in a country that sees virtue in importing inferior quality foreign coal, anything can, and might, happen. With Tarzan swinging from branch to branch in the Tory jungle it does not augur well for our old national underground forests. Under the present system all of us in this area receive our coal from Staffordshire, while the Point of Ayr coal is crushed to smithereens and sent to a massive power station far away in England. The words 'politicians' and 'logic' should not be part of the same language.

The pit's name has created a deep interest always. Point of Ayr in English, *Y Parlwr Du* (Black Parlour) in Welsh. The explanation for the English version is quite 'straight direct' as dear Amy Lloyd used to say. The name of the point of land between Talacre and Ffynnongroyw is 'Ayr Point' or the 'Point of Ayr'. No taxing of any one's I.Q. to appreciate this. The Welsh version has caused much more head-scratching.

'The name is not *Parlwr Du*, Einion bach, but *Palwr Du.'* Not Black Parlour, but Black Digger. That was the explanation I received years ago from Thomas Jones (Tommy Epworth) who had made a bit of a study of place-names. Had he not chosen with care the name Epworth on his house in respect of its connection with the Wesley brothers, his heroes? Thomas Jones named the house, the miners were responsible for grafting the name on the man from Tan-lan and his family.

'This old pit was never a parlour for your father or me, Einion bach, it has been an old back-kitchen all along. No, here's the explanation for you lad. When they were sinking this shaft they had a massive black machine doing some of the work. The Black Digger. I never went to the County School but I know that that translates into Palwr Du in Welsh. That's how the pit got its name. Remember me to your father.'

Later, I heard Rhys Jones (the musician) offering an explanation which he had heard. Some years ago some miners who had been working in Bettisfield Colliery, Bagillt, came to work in our pit. They were pleasantly surprised on coming to work in a pit where the lowest seam had a headroom of five foot-six inches in the coal-face, and in the Durbog a headroom of between nine and ten feet. The seams in Bettisfield were low and steep. On his first shift in his new pit one of the miners remarked 'This pit is a parlour compared to Bettisfield.'

I will never argue with Rhys on a point he makes in the music world. We all know that he knows perfectly well how many hairs Rachmaninov had on his head, and how many times, in his tempestuous life, Rimsky-Korsakov broke wind in the minor and major keys. I also know that the working conditions in Bettisfield were harder than at Point of Ayr. The story still circulates about Will and Gwilym Bagillt coming to the Point from Bettisfield, and on their first shift started filling the coal into the tubs (drams) on their knees as was their custom, until Tommy Lloyd (David Lloyd's brother) thundered: 'You're not in a prayer meeting now lads bach. D'ye remember how the sergeant-major used to bawl "Stand by your beds." Well, it's "Stand by your tubs" here lads bach. Ho! Ho! Ho!'

However, I must disagree with Tommy Ep and Rhys regarding the origin of the name. A few years ago I went to the Archives in Hawarden on some research or other, and saw there an old map of this area dating back at least to the middle of the last century. The name on the point of land between Talacre and Ffynnongroyw was 'Parlwr Du Corner'. This map was drawn many years before any Black Digger came anywhere near the place. Therefore the pit was named after the piece of land. Amen!

But how did Parlwr Du Corner originate? I am 'no rival to Sir Ifor' regarding place names, but I have a personal theory about the origins of this provocative name. The pit stands near the estuary of the River Dee, where a Pearl Bank is situated. The pearl bank could be reached quite easily in a boat, and some two or three miners did this long ago, scratching for the oysters and with a lot of luck they would occasionally find a pearl, which they would wrap in a handkerchief

Ned and Dafydd Jones (Picton Yard Farm) enjoying a rest (the collier's haunch). Ned was killed in an accident at the Point of Ayr on 11th June 1930 when he was 39 years old.

and take to Hughes the Jewellers shop in Prestatyn where they obtained a little money in exchange.

In old maps the word 'Dee' is often spelt 'Du'. The Welsh word for pearls is *perlau*, therefore 'Pearl of the Du' translated into *Perlau'r Du*. Is it too far-fetched to imagine that Perlau'r Du could become Parlwr Du? I don't know, but I like the idea. If there has been uncertainty about the origin of the pit's name, there is no uncertainty whatsoever about the importance of the pit locally. As soon as the lads attained the age of fourteen they would abandon the short-trousers, changing them for long-legged trousers, and follow their fathers' footsteps to the family pit. It was essential, of course, to tie the trousers below the knees with 'boings'.

'Why d'ye wear boiongs, Dad?'

'To stop the dust coming into my eyes, lad.'

There was only about a mile separating our house from the pit, and therefore I was conversant with the pit vocabulary since my childhood. The miners were always ready to have a break and chat, usually choosing the comfortable 'colliers haunch' position. Although I did not understand the terms, I heard the following terms hundreds of times—Two-yards, Durbog, Stonecoal, sling, Five-Quarters, iron-man, stone-dust, tallies, scotches, jig, sight-rest, chew-of-baccy—in fact all the miners' vocabulary apart from the swear-words. It was not often that they would swear when the children were within earshot, although some of the colourful adjectives would emerge during the hard work of harvest-time. It was during those periods that:

Picton's sky of azure blue
would take on a darker hue.

I knew about thc underground locations and the tools, and also the names of some of the men. Some names were mentioned in conversation far more often than others—Jack Garreg Lwyd, Will Glên, Hugh'r Slip, Dac Ellis, Jack y White, Wilf Two Foot and Tommy Butterfly. The latter received his nickname because he wore a butterfly collar and dickie-bow tie. Hugh'r Slip was known thus after a pub of that name (locally) in Gwespyr. Poor old Hugh's Welsh was not an example to anyone, indeed it was quite as shocking as some of today's Welsh television presenters.

Apart from carrying the pit dust and fumes home with him every day, the miner would also carry some of the pit tales to his attentive wife, and if it entailed a tale about a row, the wife would side with her husband every time. Sara Pen-y-Glasdir always listened to the detailed and colourful reports brought by her husband Pit-y-Coed. A spade was always a spade to Pit until he fell over one. On the reports Sara received there wasn't an inkling of censorship whatsoever; she accepted them all, word for word, swear-word for swear-word.

The reports which Mam got from my father at the breakfast table were delivered more leisurely, and with many fewer adjectives.

'Harry Parry Gwespyr's lad has done heckishly well. He's passed the Scholarship. This is the second of his lads to go to the County School. If poor old Harry had as much in his pockets as the lads have in their heads, he'd be a millionaire. I hope that our lads will make a mark of it when they're old enough to sit the exam.'

'They're sure to do their best, Dad.'

'I hope so, indeed, because I don't want them to come down the pit. That's why you and I have struggled to buy them those books. What are the books called, tell me?'

'Arthur Mee's Encyclopaedia, Dad.'

'Yes, that's right, but I don't see them using them very often. I see Tiw reading them sometimes, but as for this big lad I'm sure his brains are in his feet. At least, it's only on the football field he seems to make a mark of things. But he'll see how much help Dixie Dean will give him when he tries the Scholarship.'

'Yes, but fair play, Dad, remember "All work and no play makes Jack a dull boy".'

'I knew you'd stick up for him, but remember, Lal bach, "All play and no work makes Jack a dull boy" as well.'

'They can both now sing the hymns from the hymn-book without following my finger, Dad, and they both recited the twenty-third Psalm in the Sunday School meeting about a month ago, and Thomas Edwards, Tŷ Capel, Mostyn told me afterwards that he had never heard a clearer recitation of that Psalm.'

'Mam, my dearest loved one, when I was twelve years old I knew the Longest Psalm by memory, with its Lamed, Resh, Samech, Schin, and its whole lot, and I received a prize of half-a-crown from the old Tomos Roberts, Gwespyr but that didn't stop me from going to the pit. I could also answer all the questions from the "Mother's Gift", and I knew most of the "Young Members' Guide-book". Even though I knew a lot about the Acts of the Apostles, it was to the Coal-Mines Acts which I had to bow my head, Lal bach.'

'But you're a smashing father to these children.'

'You're also a smashing mother to them, and the best wife in the world; but I don't want them slaving all their lives. Imagine raising your cap to a monkey-on-a-stick of a manager like that Mr Young. Have you ever heard of a real miner going down pit in his shoes? His answers at all times are abrupt to everyone. No wonder the lads on the surface have christened him ''Young Hitler''. Take that crinky under-manager Jim Davies we've got, him and his *lemyn-cêli* smile. He's a fly risen from the horse-muck, the sarcastic scoundrel. He would never have had the job only that his uncle had been an accountant for Mr Batters, the owner. There are two big wheels above the pit, but there are many wheels within wheels around the place. I'd like to meet Jim Davies one dark night when I'm crossing ''Pont y Gwtar'' to give him a thump under his chops until his soul was marbles, and I'd relish swinging my clog to his back-side until there was a fall in his return airway.

'Oh, Dad, don't talk like that.'

'It was a kick in the back-side he got when he went to work in that pit in Huyton Quarry in Lancashire a few years ago. He thinks no one here knows about that incident, but Ned-y-Trap and John Morris Yr Ardd Ddu worked in a nearby pit, and I got the details from Ned-y-Trap, yes indeed, chapter and verse as Ned would say.'

'But Jim Davies cannot be all that bad, Dad, he's a very important member of the Independent Chapel in Gelli Bant.'

'Yes, I know, and as far as I remember the first disciples, even, were a pretty mixed bunch. There's many a one had a sudden promotion in the pit after changing his religious shirt and moving to the 'Senters in Gelli Bant. Will Cidna-bêns was telling me that I would stand a better chance of being promoted a fireman if I also changed my shirt. But I'll never go to that chapel even if it means I never get a small lamp—at least, not while J.D.'s there. I'd rather be a pagan.'

'Hush, Dad, in case the children hear you. They're getting up.'

'What's for breakfast, Mam?'

'Bacon and egg, after you've washed. And remember to wash behind your ears.'

'Ew, this breakfast's good.'

'Well of course it is. It was I who fattened the pig. It was I who killed it. It was I who salted it. It was I . . .'

'Pass the chalk here, Dad, for me to put a big ''I'' on your chest.'

'We're going to school now. So long, Mam, so long, Dad.'

'Do your best, lads bach.'

I Scan through My Days at School

Who killed Cock-Robin?
'I,' said the sparrow
'with my bow and arrow,
I killed Cock-Robin.'

Who saw him die?
'I,' said the fly
'with my little eye,
I saw him die.'

Who caught his blood?
'I,' said the fish
'with my little dish,
I caught his blood.'

And on, and on, for fifteen blood-curdling, sad verses.

The poor robin lay dead in the snow, a long arrow in its chest, large drops of blood here and there, and the other birds around. One of their number had been responsible for the slaughter, but they all did their best to have a colourful burial for the Robin.

That was one of the pictures on the wall in Miss Parry, Dyserth's classroom in Picton School, when I went there as a three year old crew-cut infant. My father was the shearer for the two lads in our house. He is the only one, to my knowledge, who did this work seated, and we had to stand in front of him, while he crossed his legs like a vice around us. It was impossible to move an inch until the performance was over. My father had never heard of 'short back and sides'. He only had one style,—all off. According to his philosophy the head would be easier to wash, and a warm shelter for the fleas would be destroyed.

I cannot recall the picture of Cock-Robin giving me a great

Picton School.

fright at the time—a little sadness, maybe. I shudder much more now when I think of it sixty years later. The 'Hall' was graced by pictures of the important War Generals,—the British ones of course. These were the Great War heroes who had come to fame about twelve years earlier. These were the custodians of the British Empire. There was also in the Hall a large map of the world, with its huge red areas to denote the British Empire. It was, of course, after some years that I realised how appropriate and significant that red colour was. In spite of Cock-Robin, the warring Generals, and the red map I enjoyed undisguised happiness there.

The school consisted of four class-rooms, and four teachers, namely, Miss Parry, Dyserth with the infants, Miss Parry, Picton in the next room, then Mr Elwyn Morris, and the Headmaster Mr J.B. Thomas, Gwespyr, a native of Llanybydder, South Wales. Christened John Bydder Thomas, but to all, young and old, he was J.B.,—behind his back of course. This Will Hay's main goal was to teach 'copperplate writing' and 'the parts of a pound' to us all.

'One half-penny is one-four-hundred and eightieth of a pound, one penny is one-two-hundred and fortieth of a pound . . .' and on, and on. Before long Miss Parry, Picton married Glasdir Edwards, and quite naturally gave up her work at school. Burning brassieres was not the most esteemed experience in those days. They possessed fewer luxuries in their houses, but their offsprings were never latchkey children. She was replaced by Miss Pearson, Ffynnongroyw, until she, in a few years time married and left. In a while the staff was increased when Miss Proffitt arrived. She was a dear soul, full of tenderness, and had an excellent gift for telling stories. When I reached the top class the boys would often become very noisy, and quietness would not be restored until Miss Proffitt told us a story. I recall three student teachers during my time—Charles Williams, my cousin, Albert Henry Parry, and Charlie Parry, the latter two, of course, being the sons of Harry Parry, Gwespyr, who worked down pit with my father; these were the lads who were 'doing heckishly well' at the time. Dewi Pearson, and Dewi Lloyd Jones also taught there for a few weeks only.

All the teachers were fluent Welsh speakers, as also were the vast majority of the children, but, as in all schools, our subjects were taught in English. Although we greeted God in the morning as 'Our Father we charge in Heaven', I believe we used the Welsh version sometimes. Many English songs were committed to memory, 'Rule Britannia' and 'The Grand Old Duke of York'. We also learnt many pieces of English poetry, and fair play to the teachers, they did this work quite thoroughly, because some of these poems are still embedded securely in my memory sixty years later.

I remember, I remember the house where I was born,
the little window where the sun came peeping in at morn.
It never came a wink too soon, nor brought too long a day,
But now I often wish the sun had borne my breath away.

We learnt 'The Daffodils' by Wordsworth, and sections of Hiawatha, and also 'The Wreck of the Hesperus'.

It was the schooner Hesperus
which sailed the wintry sea,
and the Skipper had taken his little daughter
To bear him company . . .

A huge storm rises, and the little girl is drowned. This poem impressed me deeply. It would make me immensely sad, but I returned to read it time after time. A father losing his only daughter . . . The poem 'Abu Ben Adam' was also seared on my memory.

In July, 1983 I met Mr Elwyn Smith, the Headmaster who had succeeded Mr J.B. Thomas in Picton School. It was he who had taught us many of these poems. Quite naturally our conversation turned to poetry, and to the poems he had taught us at school. Special reference was made of 'Abu Ben Adam', and although we were in the Kwik Save Supermarket's car park in Prestatyn we recited it together loudly, caring nothing for those who passed. Both master and pupil had a good memory.

Although I pride myself on having a good memory I cannot say with certainty that we were not taught Welsh poems in Picton School, and I would not like the teachers to suffer any injustice. Still, I believe that Welsh material should remain better than English material in someone's memory who once wrote, 'A stitch in time saves NAIN' (granny!)

When Charlie Parry (the student teacher) saw this in my book, he laughed loudly. I failed to see anything at all amusing, and said under my breath . . .

> Charlie, Charlie, chuck chuck chuck
> went to bed with three chuck chucks.
>   One died
>   the other cried
> Charlie, Charlie, chuck chuck chuck.

Mam never experienced any difficulty in getting me to school. In spite of that, I well remember having an occasional 'shift to the band' in the early days to go with my father to fetch coal from the Coal-Yard in Talacre. Although the miners were on a low wage, they had a good load of concessionary coal each month very cheaply. Some of the carters like John Jones y Weit would convey the coal for the miners in their carts. In order to save a little money, my father would fetch his own coal. He would borrow a horse and cart from Dafydd Jones, because they had an unbroken agreement for years that if my father gave a hand at harvest-time then he could borrow the horse and cart each month. Occasionally I was allowed to accompany my father to the Coal Yard. Why, I do not know, because I was far too young to be of any help. Maybe the intention was to give me a little treat. Quite early after breakfast I would go with my Father to Cae Bryn to fetch the mare 'Pet', or the horse 'Tyrffi' to the stable, help him with the traces, and then hook the horse to the cart. We would make for the Coal-Yard, not forgetting to squat right down on the floor of the cart as we passed the school, lest the Master should see. A further quarter of a mile near Gelli Bant stood 'The Bluebell Mine' where three or four ex-miners from the Point of Ayr had sunk a drift mine. They succeeded in striking quite a good seam, indeed because there was such an abundance of coal in this area, it would be harder to fail to reach a seam! Plenty of coal here in the Gelli Bant, and yet my Father would go for another two miles. To me it seemed senseless. At the time the word 'concessionary' was not in my vocabulary.

After arriving at the Yard the horse and cart were weighed on the large weigh-bridge, before going to the large waggons for the cart to be filled. When my father believed that the load was near to his allowance we would return to the weigh-bridge, where Eseiah Pearson would be waiting to give his verdict. Most times the genial deacon from Moriah Chapel, Ffynnongroyw, would agree that the load was quite hefty but not too far over the mark. Occasionally, if the load was very excessive, some of the coal would be thrown down near the door, where it would come in useful to burn in the Weigh-bridge office. We would both climb into the cart, sit on the load, and make for home. Sometimes I would receive a penny from my father, and I would call at the little shop at the bottom of the Rhiw, Gwespyr, to buy ten caramels. On the way home the chewing would be superb, one with his caramels, the other with his twist-baccy. 'Gee-up, Tyrffi.'

I do not believe that my visits to the Coal-Yard were responsible for one frighteningly short composition of mine once. It was during the new Headmaster's early days at school, when I was about ten years old. The title appeared on the black-board,

'What I would like to be when I Grow Up', and this was my innocent and honest response . . .

'When I grow up I would like to be a teacher, in order to have an easy job, plenty of money, and long holidays.' No more, no less. I could not think of any more to say. What need to say any more? However, the Headmaster succeeded in finding much more to say, and he insisted on 'having his say' for half the afternoon. He strode backwards and forwards, scolding without mercy.

In less than a year's time a composition I wrote must have pleased the examiner quite a bit more, because in no time, one of Alf Evans's lads followed Harry Parry's lads to the County School in Holywell. Arthur Mee's (hardly opened!) Encyclopaedia, The Hymn Book, The Psalms, old J.B.'s 'parts of the pound', the other teachers' toil, had done their work, and furthermore, I trod the narrow road that would keep me from Point of Ayr's hell. At least, that was the general belief at the time.

When I attended Picton School I knew that the teachers were fluent Welsh speakers, although English was the language of the classroom. Not so in Holywell; indeed it was only after I left that school that I realised that many of them were Welsh-speaking. The Welsh master, Gwilym Thomas always spoke to us (the Welsh-Welsh) in our mother-tongue, although his partner Miss Jones, who taught us Welsh Literature, spoke to us, and taught us through the medium of English. The Headmaster, Rhys T. Davies spoke to us occasionally in Welsh, and he made sure that one morning's service each week would be conducted in Welsh. He also ensured that the discordant School Orchestra should partake in one service each week. He must have believed in the portion of scripture which says, 'Make a joyful noise unto the

The Upper Classes in Picton School *c*. 1937. The author, with his head inclined towards the Student Teacher Charlie Parry on the left. Elwyn Smith the headmaster is on the right of the photograph.

Lord!' The remainder of the Welsh language I heard would not fill one side of a record by Dafydd Iwan.

The walls near the Headmaster's office were adorned by a photograph of one of the previous Headmasters, J.M. Edwards (brother to O.M. Edwards) and that of Emlyn Williams, an ex-pupil, and not with photographs of warring Generals like those on the wall in the Primary School. The Headmaster (the Boss) was a Christian and a pacifist, and this, I am sure, would have influenced his choice. However, the connection with the war could not be completely severed because Lieutenant-Colonel J. Llewelyn Williams was Chairman of the Governors. In spite of his pedigree he appeared to be quite gentle and meek in his old age, and occasionally he would persuade the Headmaster to grant us an extra half-day's holiday when someone in the school had excelled academically.

It was on the football field that I had the most enjoyable experiences of those days, but also one experience which hurt me. The pain remains, and although, over the years, I have tried to erase the memory, it remains in the background. In order to play football it was necessary to have shorts, and at the time Mam could not afford to buy me a pair. My sister, Evelyn, wore shorts during her games lessons. Therefore it was only reasonable that I should borrow them. The difference in the shorts worn by the different sexes was trivial, the ones worn by the girls fastened with buttons in the top, and the ones the boys wore had elastic around the waist. The difference was barely noticeable. Nevertheless one lad from Flint noticed the difference, and I was the target of his mockery, which pierced me as painfully as a rusty sword. I'm sure he completely forgot about the incident, but the scratches remain in my memory to this day.

I was not consoled either by the fact that I was the only one in school at that time who wore heavy hob-nailed boots. When Tiw, my brother, followed me in a couple of years, at least there were two of us keeping a noise like cart-horses along the corridors. Because we had to travel down the Brêc footpath to catch the bus every morning, my father insisted that shoes would be useless to travel through the mud. Although I was not conversant at the time with the term Hobson's choice, that was what I got.

Wearing the heavy boots was 'not wholly bad' (like Eli Jenkins's congregation!) In spite of their weight I could run in them quite quickly. When I took them off and wore the football boots, I did not know that I had anything on my feet, and I could rocket along the field. I was mocked for wearing Evelyn's shorts. My heavy boots were also cause for mirth. No one ever mocked me when I played football. On the left wing, or in the centre forward position I mocked everyone, especially the goalkeepers! I was master of them all.

For the first three years of my life I was left-handed; with my left hand I knocked scores of nails into the big table; this was also the hand used for throwing the dice whilst playing snakes and ladders. When I went to school I was taught to write with the right hand, contrary to my inclinations. That was the system, but I'm perfectly sure that this alters the personality. Something must be responsible for my satirical poems. I still write with the right hand, as taught, but the inclination is still in me, to do many things with the 'southpaw's piledriver'.

Like the big hob-nailed boots, the curse of the hand-change was a blessing for me on the playing field. There, my left foot was allowed its perfect freedom, and I could express myself without any fetters on the body or mind. Recently, when I met the Rev. E.H. Griffiths, Rhyl, we naturally talked about the time we had been playing together in Holywell years ago. One of his comments was:

'You had a deadly left foot,—and a dead right foot.'

All the teachers noticed my dexterity with the ball. I well remember Mr Christopher, who tried to teach us Latin, telling me once, as he returned my rather disappointing exam paper,

'Ainion.'

(That was another problem which raised its head quite

often: only the chosen few could pronounce my name correctly; I was called Ainion, Enion, Eenion, and sometimes Onion).

'Ainion, the next time you sit a Latin exam I suggest you come in your football boots. You always seem to do well in them.'

Amo, amas, amat? To score meant much more for me. I played for the 2nd XI during my second year. For anyone possessing a little self-respect this is not a team to linger in. Indeed I would go as far as to say that being made its captain is more of a condemnation than an honour. Since I started in Holywell I would look at the notice-board every Wednesday afternoon when the names of the 1st and 2nd XI teams were displayed in readiness for the oncoming Saturday. One Wednesday afternoon when I had just started on my third year, the name I saw in the inside-right position was E. Evans. The notice was definitely official, because it had been signed by S. Linton Rees, the teacher in charge of the 1st XI, and also by the captain, Frazer Phillips from Bagillt. Was it possible that I could be that E. Evans? Frazer, the ginger-head gentle-giant had to be found.

'Frazer, who is this E. Evans who's playing for the 1st XI on Saturday?'

'You lad, then we'll have another copper-knob in the team. We already have Gwyn Morris and myself. Are you available?'

Available indeed! Being allowed to wear the red and black quartered jerseys, being a member of the same team as Frazer Phillips, Billy Roberts, Howell Jones, Gwyn Morris and Ronnie Jones. Inside right was not my best position but what difference, I would be content to play in any position, indeed I would gladly be a corner post if the flag on it was red and black. In my youth this was more precious in my sight than the riches of Peru, and with a smile as wide as two football pitches I returned to the same world as Frazer.

'Yes, Frazer, I'm available.'

At long last the Saturday dawned, and the home game against Mold. I know we won, I know also that I scored one of the goals. I wandered a little to the right wing, received the ball, beat two opponents, and although I was a considerable distance from the goal I blasted the ball, and that sped into the goal beyond the outstretched arms of one of Wil Bryan's descendants. Yes, the deadly left foot had been 'true to nature'. I was surrounded by my excited team-mates, but of course the greatest thrill was Frazer's hand-shake and comment,

'Good for you, copper-knob.'

I 'scored a legion of goals' (as Mr Rees recalled years later, in his retirement party) but this was the sweetest of all. I could go back to the field in Holywell (indeed I did so lately when I was being honoured by the Old School for winning the National Eisteddfod Chair) and stand on the exact spot from where I struck my first goal for the 1st XI.

Mr Rees the Chemistry master was our referee. He was lovingly christened Sidney Linton Rees. Being born in Llansamlet was sufficient reason for him to be called Sammy Rees through his years in Holywell. It was he who pumped the footballs, and also accompanied us on the Crosville buses to our away matches. If the bus was a double decker, the teacher would travel on the lower deck at all times, leaving the team to travel on the upper deck. Well he knew that the stains on our fingers had not been obtained collecting conkers! He rendered years of voluntary service like this beyond his daily duties in the class-room. Thank you, Sir.

Dafydd Iwan has issued a record regarding the school lessons he received. The story during my time was very similar. There were three classes for each year, A, B and C. The barometer-finger of my education stuck in the centre of education's Liberal Democratic B stream, no sun shone to lift me 'on the wings of dawn' to the A stream, neither did a wind blow to abase me to 'make my bed' in the C stream. As I have previously mentioned Latin has always been Latin to me, some utter darkness; but I saw Welsh as 'the fair light of break of day'. When the day of reckoning in my case arrived at the end of five years, I gave some of my teachers a pleasant surprise—I scored at the desk. Nothing stunning of course (hat-tricks not matrics for me) but enough for me to return to Picton School at a student teacher to prepare for entry to Bangor Normal College. But woe-betide any lad who becomes eighteen when his country is baying to use him in its bloody extravaganza.

With much more cruelty than was seen in the picture of Cock-Robin, the German eagles, and the British birds of prey were attacking. And like the killers in the picture they enjoyed arranging colourful funerals, with full military honours. The comparison is perfect—the innocent dying, and the most innocent of all mourning and longing.

> Who'll be chief mourner?
> 'I,' said the dove
> 'I mourn for my love,
> 'I'll be chief mourner.'

A mother wept on the banks of the Thames
and a mother wept on the banks of the Rhine.

Snappin With the Bevin Boys

We had to join, we had to join,
We had to join old Bevin's Army.
Fifty-bob a week, nothing much to eat,
hob-nailed boots and blisters on your feet.
We had to join, we had to join,
We had to join old Bevin's Army.
  Bevin the rascal.

That was the song I heard when I went to Swinton Colliery, Manchester at the beginning of 1945 as one of hundreds of Bevin Boys. At that time the Government placed an importance on the coal-mining industry, and in order to ensure more miners, a small percentage of the eighteen-year-old lads were compelled to go to work in the pits. Because this arrangement was brought about by Ernest Bevin, the boys were christened Bevin Boys. I chose to be one of them because I am a pacifist. As far as I can see, the commandment, 'thou shalt not kill' is emphatic and final.

'Thou shalt not kill'—Moses or Pharoah
'Thou shalt not kill'—Lloyd George or the Kaiser
'Thou shalt not kill'—Churchill or Hitler
'Thou shalt not kill'—Thatcher or Saddam Hussain
'Thou shalt not kill'—Amen.

Maybe my pacifism can be traced to the early years of my childhood in Cartrefle. We only ever had one dog, named Toby, and he did not have a long life. We had many cats, and occasionally, after one had gone fat without over-eating, kittens were seen around the place. It was always my father's work to drown them. Once, when I was about seven years old, it was decided (by Cartrefle's law-maker) that I should assist in the task. Four kittens were born in the Old Hut, and when the mother turned her back, my father brought them out to the garden and drowned three of them, one by one, in a jam-jar.

'You drown the next one.'

I put her in the water and she died quickly. I'm still guilty. I'm still a pacifist. I am still a man repulsed at lifting a gun. Lady Macbeth is not the only one to have the guilt of death on the blood-stained hand. 'Out damned spot'.

In Manchester I found myself amongst many lads who had never seen a coal-mine in their lives. Strange also to them were the thick, striped flannel vests and drawers that the real miners wore; and when they saw me wearing them on the first morning one of them asked me, 'Are you going to a Fancy Dress Ball, Taff?' We camped in old army huts; and the first lad I saw was Clifford Tŷ Capel, Ffynnongroyw. This eased a little on the longing felt by both of us, and we shared much of each other's company for the month which the course lasted. Two other good mates I made there were Cecil Evans and Armon Ellis both from Llanarmon, the latter subsequently becoming a high-ranking police officer. Manchester has received a rather special name because of the abundance of rain it receives. I have no reason to doubt that the name is anything but appropriate because every morning the watchman would knock the door of each hut with his stick and shout, 'Come on lads, it's half past six and its raining like jiggery.' We went down pit some days, and other days were spent nodding through lectures, or sleeping through coal-mining films. After all, none of us were used to a 6.30 a.m. disturbance.

It was quite a pleasant time, although I had a fright during one of my first evenings when one of the English lads came to me, enquiring as coolly as if he needed directions to the Post Office,

'Where's the nearest red lamp, Taff?'

I kept well away from him for the remainder of my stay in Manchester. Better a dose of longing than a dose of V.D.

At the end of the month we would all receive a ticket informing us of the pit at which we were to commence work the following week. Cliff received a ticket to go to Point of Ayr. I received one to go to Black Park Colliery, Chirk. I nearly died of shock. My home was only a chew-baccy spit from the Point of Ayr, and these monkeys were going to send me miles and miles away from home. I tried to reason with the officials about the folly of the thing, but it was like knocking your head against the coal-face; their instructions had been received from Newcastle.

I arrived home in a big panic, and although Mam had a bumper meal ready for me consisting of potatoes, home-cured ham and eggs, farm-butter butties, and an apple tart and custard, I did not enjoy the food half as much as I usually did. Mam and my Father were also very disappointed that I had to go so far away from home to work. 'Finish your food, and we'll go to see Jack Garreg Lwyd' was my Father's verdict. Jack was the local delegate of the N.U.M. who lived

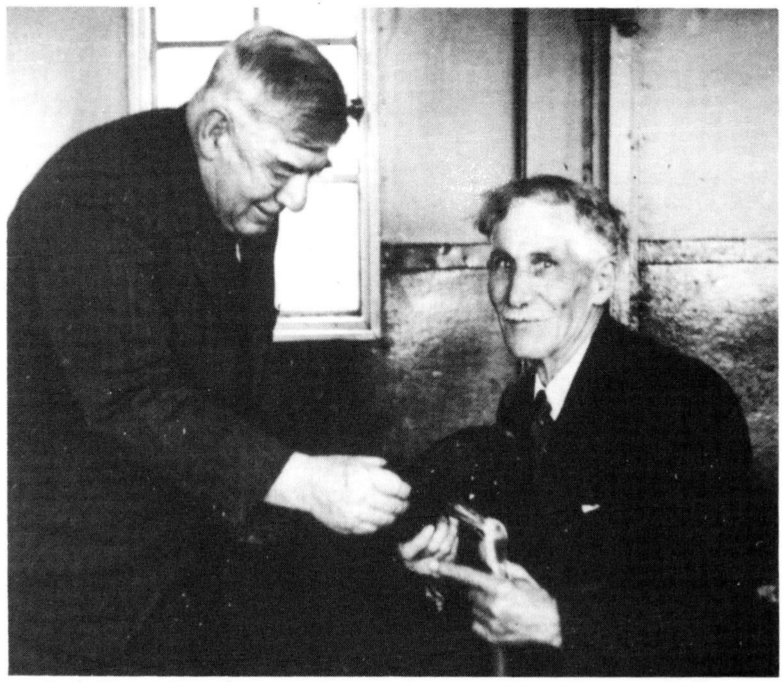

Jack Garreg Lwyd presenting a retirement gift to Barnet y Sadler.

in Trelawnyd. A friend took us there in his car, and luckily, the contemporary Abu Ben Adam was at home.

Jack and my father had attended school together. It was they who fetched a bucketful of buttermilk from a nearby farm for the headmaster each week. One day because the churning had not been completed, the farmer told the lads to return in about an hour. In this farm, as was customary in those days, a large tub was situated in the yard to hold sour milk and swill for the pigs and calves. The lads gave each other a knowing wink, and it was not necessary for them to return in about an hour.

'Well, come in, lads bach.'

That was the greeting we received in a voice which was a cross between Paul Robeson's and William Parry, Picton Hall's bull.

'And what can I do for you?' was the question before we were seated.

Fair play for Jack, he was always doing something for someone.

Doing his best to re-establish the Union in the pit. At one time the N.U.M. was barred at the Point of Ayr. The local Industrial Union was formed by the Manager and his suck-lings—a union of the management, by the management, for the management.

It was Garreg Lwyd, Jack Grffiths and Moi Owens who were responsible for bringing the N.U.M. back. Working in secret at first. Holding their meetings in Dafydd Isfryn's coal-shed in Tan-lan, and Mary, Dafydd's wife giving them a cup of tea. Moi collecting the weekly subscriptions on a little table on the Main Road in Tan Lan. For fear of reprisals at the pit some of the miners would leave their contributions at a drapers shop in Ffynnongroyw, owned by John Roberts, or with another John Roberts, Rock-houses, an insurance collector to be collected later by Moi. As the membership increased, so also did the confidence, and the little table would be moved nearer the pit, up to the Railway Bridge, then to a hut on the colliery yard. By now, the N.U.M. office is only a short-rail's length from the manager's office. Doing his best for everyone, doing his best to get pit-head baths, doing his best to get the 'skilled shilling', more 'waiting time' and 'water money' for the men. Doing his best for everyone without feathering his own nest.

'This lad's had orders to go to work in Black Park, Chirk on Monday.'

'Well, 'pon my soul, what next, tell me?'

'What will we do Jack?'

'Wait a minute. Listen, is your health alright?'

'Ew, yes, smashing, Mr Jones.'

'Yes, you've got a pretty good grain, your mother must have put quite a lot of Indian Meal in your trough. No, I was thinking is it possible for you to have a pain in your back by Monday, and go to the Doctor for a paper.'

'Fine, I understand now. I'll go to the surgery in Ffynnongroyw, and if the Doctor isn't there I'll go to Prestatyn. I think my back is starting to hurt now, Mr Jones.'

'Chip off the old block here, my old Alf. Yes, you get a Doctor's paper and send it to Black Park, and then I'll do my best. Remember, I can't promise anything, but I'll do my best.'

'Thank you very much, Mr Jones!'

'Is there enough sugar in that tea, Alf?'

'Champion, thanks, Jack.'

'I've got a rather important meeting on Wednesday in the Miners' Institute in Rhos. We're going to see plans for us to try to get Pit-head Baths to this place; it's a crying shame that we're the only pit in North Wales compelled to bring the coal-dust to our houses. It's no use our wives getting the place spick and span every day. The minute we come home the house is like "Dusty Canyon". But I know alright why those in the Head Office are reluctant to help us to get any improvements, they haven't forgiven us for that ups-a-daisy Industrial Union, that was yonder, although neither of us belonged to it. It would be hard for us to look this lad in the eye now, Alf, if we had gone to that Tribunal in London with the manager and his sucklings, to say that everyone was happy in the Family Pit. I know we carried the butter-milk and butter for the Master, but we never once buttered the manager. No, never, thank the Lord.'

'So long, now.'

'So long, Jack.'

'Goodnight, Mr Jones, and thank you very much.'

I was the first in Dr Armstrong's surgery on Monday afternoon.

'And what's the matter with you, Frank?' was her first question. She always had difficulty in differentiating between me and Frank Parry, Cyrnol, both of us the same age, of similar build, and fairly similar features, to someone who did not see us often.

'I'm not Frank, Doctor.'

'Oh no, of course not, you're Alfred Evans's lad from Picton, aren't you?'

The name Einion was a bit of a stumbling block to this one too.

'So what's troubling you then?'

'I've hurt my back, Doctor.'

'And how did you do that?'

'Playing football on Saturday, Doctor, and it's very painful,' I said with a politician's dishonesty.

'Are you working?'

'No, Doctor, I could hardly walk this morning.'

The second lie slipped over my tongue as smoothly as a tub slides over the greasers.

'I'll give you a prescription for some embrocation and a Medical Certificate for a week.'

'Thank you very much, Doctor.'

I made a quick sharp exit, carrying the doctor's paper in my hand as proudly as Moses carried the Ten Commandment Tablets from another 'waiting-room' on the slopes of Mount Sinai. The embrocation would also be smashing for my legs before the match on Saturday. By Wednesday, Jack Garreg Lwyd's best had been good enough. The note left on my

Father's lamp in the lamproom explained that Jack had spoken to a Mr Rodgers, of Wrexham, and had made the necessary arrangements for me to get a lamp in the Point of Ayr on the following Monday—provided my back was better of course! Apart from the month's training in Manchester another fortnight was needed in the Point of Ayr. The Training Officer was Elias Hughes, Mostyn (Lei Top), and Clifford and I were quite obedient trainees to him. By this time my father had been promoted to a fireman, and when my training was completed, he asked permission for me to go to work in his district on the night shift. At that time this was the repair shift, while all the production was done on the morning and afternoon shifts. The duties on the night shift were setting additional props, levelling the bottom when that had become swollen, improving the air-flow by fixing brattice cloth in suitable places, scattering white stone-dust in the roadways, very like the old 'sower who went out to sow' in the Bible (The stone-dust would cling to the coal-dust and make an explosion less likely). These duties were carried out conscientiously by the old men, but it was leisurely. Although I was quite happy on this shift I was plagued by a desire to go to work in the coal-face, to go with the lads on the hooking-shift, to prove to myself and to others that the years at the County School had not softened me.

In no time my wish was granted and I found myself driving an engine, hooking clippers, driving outside piece, driving in the face, handling scotches, riding the horses tail-chain, a master on the chain-clip, and then the ultimate challenge—my first loading shift. One day Raymond Smith, who loaded in the South of the Two Yards with Tegla Parry, Llew, missed a shift. Because there was no experienced loader to take his place, the overman asked me to go there. I knew that it was donkey's work, but lest anyone should say that I was a fountain-pen pusher and nothing else I agreed to go. The work involved loading coal without pause into tubs, because the loaders were on piece-work,—the more tubs that went to the surface, the more the wages. The only break one had was when the fireman fired the top or the bench in the coal-face; about a quarter of an hour was taken in mid-shift to have snappin and a drink of water. On my first loading-shift I emptied my water-bottle long before snappin time, but the rest of the set made sure that they shared their water with me. We loaded three-score-and-ten tubs that shift, equalling exactly what Raymond and Tegla had loaded on the previous shift. I felt more than satisfied—and was more than ready to go to bed after reaching home.

At this time there were about fifteen Bevin Boys in the pit. The first of them was Hugh Jones, from Anglesey, who started a little before me. Quite naturally, he was christened Hugh Bevin. I don't know who christened him, but the name stuck straight away. Many of the Bevin Boys came from the Rhyl area—Danny Jones, Tiny, Les Cole, Derrick Poovah, Eric Lewis, George Yates, and Malcolm Simcock.

There are some humorous tales about some of them. In this period every pit was given a production target to aim at, some thousands of tons each week. If the target was exceeded, a flag was hoisted on the upper-most point of the winding-gear above the pit. Now, fair play, it was the Red Dragon which was hoisted to this position every time in the Point of Ayr.

The Bevin Boys entered to the midst of this atmosphere. One morning a group of us sat at the bottom of the Stone-coal sling, waiting for John Williams (Jack Smart) the overman to send us to our different tasks. In our midst sat a new Bevin Boy. John Williams noticed the stranger and assuming that he was English, like the majority of his tribe, asked him in his high-pitched screechy voice, 'And what is your name, Marro.' The answer he received to his question this time was,

'Les Cole.'

'Less coal, uffarn, it's more bloody coal we want, not less coal.'

There is also another true story from the same period about Derrick Poovah, who comes from a family of sailors in Rhyl.

Some names from this family appear often in the Daily Post. When Davy Jones is rolling his sleeves up, and many a ship is in danger on the sea, one can be assured that one of the Poovahs will venture out to assist. The same group, more or less, in the same place at the bottom of the Stone-coal sling and John Williams sending us to our different locations. When he reached Derrick Poovah, he said in his usual falsetto,

'Marro, I want you to go and drive the horse, Paddy, in the Far-end of the North.'

The stunning answer he received was,

'Listen here, John Williams, I'm a Bevin Boy, not a bloody cow-boy, and I'm not going to drive your horses, and if you stop my bonus I'll report you straight away to Jack Harry Clwyd.'

'Yes, you go to Jack Garreg Lwyd, and go to him also when you think you know enough to get the skilled shilling—you won't get it from me, not while you've got a nose on your face. Skill indeed? The only skill you've got is with the hobby-horses in the Marine Lake. What have we done to deserve some useless devils like these Bevin Boys, tell me? I don't know indeed. No indeed . . . Is John Boxer in work? Oh yes, come here John B. Will you go to drive Paddy for me?'

'O.K. John Williams.'

'Well done, Marro, I'll see you right in your wedding.'

Dear John B., innocent, obedient, unable to read or write, nor even count his wages. When he got his money-tin on a Friday afternoon he would ask someone close by,

'Is this right, Ken?'

To John B. everyone was Ken.

In spite of his educational shortcomings he had one very special gift, namely playing the spoons. He would regularly entertain the patrons in the Farmers Arms, Ffynnongroyw on Saturday nights, after obtaining the correct oil to lubricate his elbow. Someone would go to the piano—Ness Toes usually—and John B. would start to perform with his spoons. John always remained in the back-ground, but on these occasions he commanded the centre-stage. He has been in heaven now a few years. We know that there are harps there, but I believe that the Almighty God in his infinite love has given special permission for John B. to praise the Lamb with his spoons. Thus the harmony will be more complete.

Because some of the Rhyl lads had uncommon names such as Cole and Poovah, it was unnecessary for them to have nick-names, although Point of Ayr's nick-naming press, like the Lolfa, found it hard to abstain at times, and gave nick-names even to these.

Because of his loud and frequent laugh Derrick Jones became Derrick Ha-Ha. Not only was Derrick given a nick-name but it was he who minted a new nick-name for John Pentre Ffyddion, one of the local lads from Trelawnyd. The name Pentre Ffyddion was too much of a skipful for Derrick, and he was satisfied with John Pentre Pwdin. In no time at all, this was abbreviated to John Pwd. That's the name that stuck in a groove until his grave. Eric Lewis was christened Pastor—because he was quite a swearer, according to some. No, said Eric (naturally). According to him he received the nick-name because he showed an interest in an evangelical crusade in Ffynnongroyw, conducted by Pastor Jeffries. Around this time there was an unusual phenomenon in Ffynnongroyw when a group of youngsters were influenced by the English Evangelicals. The appropriate vocabulary sparked through them, and being as they had sufficient oil (Matthew 25. 4) they could move from Genesis to the Revelations in ten seconds. Some worked in the pit, and an occasional one of them carried his Bible with him. 'Hello, Brother,' was their greeting to each other when they met, even underground. I'm afraid that their experiences were of a camp-fire variety, and the brotherhood did not last for long. Not long enough for even one of them to be nick-named Wil Brother.

Courting and Loving for Life

There is seeing and seeing. Because Betty and I had been born within half a mile of each other, and were attending the same Chapel, the same Band of Hope, and the same school we must have seen each other frequently during our childhood. I was bound to have seen her playing hop-scotch with the other girls when I went occasionally to Lisi Parry's shop to buy sweets. I recall seeing her acting Little Red Riding Hood in the play. Seeing her, like I saw Vera Lewis taking the grand-mother's part in the same 'production'. I must have seen her often in Picton School, although neither of us can remember anything about each other in school. It was a general seeing. An indefinite, unemotional, grey, and level seeing.

Betty and Einion on Cae Bach Stile.

But regarding the other seeing, that is completely different. It is not some skimmed-milk seeing: this seeing is creamy. This seeing possesses electricity that Manweb knows nothing about, and a magnet stronger than the North Pole. It has a rainbow of colours, volumes of analogy, and oceans of harmony in one moment. Yes, I saw Betty.

At the time I played football in Cae Bryn. Because of the war, and with the majority of the young men in the forces, there was not much organised football. The Dyserth League had finished, and the Junior Cup was becoming tarnished in Mostyn or Anglesey. Furthermore, many a referee who had to squat under a shower of sods had to take shelter from a shower of bullets. However, a football team, to play friendly games, was formed by some of the local lads, Don Brooks, Alun, Myrtle House, Rab Choudhury, Emlyn Lewis, Ifor Price, Herbie Lewis, Dei Jones, John Morris and others. For a while everyone paid a shilling a week, and soon enough had been collected to buy two footballs and a set of jerseys from Jack Sharp's, Liverpool. Red jerseys, with a white collar. It was decided on the red colour as a mark of our admiration for the 'Old Team', the wonderful Robins. Some were quick enough to point out that the resemblance between the two teams finished with the colour of the jerseys. A similar group had been formed in Trelawnyd and we played against each other many times, courtesy of bicyle transport. Many military camps, with football teams connected to them, had been established in Kinmel, Prestatyn, and yes, believe it or not, a small military camp had been established in Picton of all places. Four or five search-lights beamed from Cae Bryn, and there were about thirty soldiers in the camp. An ample number to field a football team at any time. Strange names and strange nick-names were commonplace to us in no time. Johnny Croft, Tommy Twinkletoes, Len Camp, and Bob Myles. The latter two married local girls and lived here the rest of their days. Bob Myles graduating to Bob Kleen-eze. I was about seventeen at the time, and Betty was about thirteen. I enjoyed myself in the centre-forward position,

Penyffordd F.C. *c.*1950.
Back Row: Michael Nulty, Penri Jones, Herbie Lewis, Ernie Robinson, Heddy Lloyd, Gerald Wilson, Gwyn Jones, George P. Jones.
Front Row: Will Jones (First-aid man), John Morris, Elfed Jones, Einion Evans, David Beatty Edwards, Alf Jones.

while Betty watched from the touchline. This had probably happened often before, but this time I saw her. I cannot remember who our opponents were, I cannot remember who won the game, I cannot remember whether I scored a goal or not. But Betty scored a goal without moving a step. That moment I decided that here was my Juliet, and I vowed that this was the place where I would hang my tally. I knew that she was too young at the time for us to start courting, but I had seen Betty. Truly yes. I knew she was too young for us to start courting in twelve months time also, but that was no deterrent. A shy courtship at first, but little by little more openly, then officially.

A sort of *Aelwyd yr Urdd* was formed in Peniel Chapel Schoolroom at that time. The chapel of the 'dear brothers' but we had discovered the importance of the 'dear sisters'. A leader was appointed, and members were registered. Our football team was called Penyffordd Aelwyd. A small choir was formed, and the interest was sustained long enough to learn the song 'Hail, Smiling Morn'. A trip was organised to

Llanystumdwy to see Lloyd George's grave, and we played darts and billiards. Membership was for Welsh speakers only, but the rules were bent for Rab Choudhury, because he was a good footballer, and it was his billiard table that we used! But somehow I believe that O.M. Edwards had something more appropriate in his mind than postman's knock as the main item of the evening meeting, as announced in the Sunday School meeting long ago. Maybe it is my increasing old age which made me write that last sentence, and on reflection it can be argued that Post-man's Knock could be included in the Folk-dance section!

As one can imagine this Aelwyd proved a very popular place. I remember our leader being anxious to put a motto on the wall above where the pulpit used to be. In order to create interest, and hopefully, to get a worthy motto, he decided on making it a competition. When the result was announced our Tiw was the winner, and the motto was displayed in a place of honour for years. The adjudicator believed, 'on word and conscience' that the winning line deserved the prize mainly because it was a line of 'cynghanedd' (strict metre in poetry). Alas, I see by now, that the less said about the adjudicator's knowledge of 'cynghanedd' the better. If the harmony of 'cynghanedd' was strange in this Aelwyd, I know for a fact that 'many a chord harmonius' was struck there, and at least ten of us were joined in marriage. Olwen and Dyfrig, Gwladys and Alun, Eluned and Donald, Gwen and Tiw, Betty and myself.

Because we started courting at a very young age, our courtship was a long one, and I'm sure ours followed the same pattern as most couples of our times. By today, many couples sleep together before saying Hello. It was only after a considerable time that courting couples of our day went to each other's houses. We were travellers of the highways and meadows, until there were concrete signs that we were taking the matter seriously.

Betty and I met on alternate evenings, at seven o'clock outside her house. The programme, *Dick Barton, Special*

Agent was broadcast on the wireless at that time, and when the programme ended Betty would look out of the parlour window to see if her 'special agent' had arrived. In the darkness of winter I pulled on my cigarette for her to see that it was me. At the time I was a very heavy smoker, but about twelve months before I was married I decided that the money spent on cigarettes was better in my pocket than in the pockets of W.D. and H.O. Wills, and I have never smoked since. Although I was such a heavy smoker, I never once smoked in front of my father.

Our arm-in-arm journeys hardly ever varied their pattern. We would start towards Picton, travel up to the school, past Ffordd Ddŵr and Maes Mari to Glan'rafon, then past Bryn Mawr, Tan y Bryn, Gwynfa Chapel, and then back to Bryn Teg for the last kiss. Of course, the stations on the journey were numerous, Pen-y-Glasdir gate, Picton School corner, Maes Mari's fence, Cae Bach and Cae Dafydd stiles, an occasional gable-end, and the shelter of an occasional heavily foliaged tree. Because we were so familiar with these stations it mattered not which way the wind blew, we had a suitable shelter. The big night of course was Saturday, when we went to the pictures in Rhyl. When the football team played at home we went together on the bus, but when we had away fixtures we would meet in the Crosville Bus Station, under the clock. Couples awaiting Cupid. The girls would wear their best clothes and their precious nylon stockings which they had often received from their boyfriends. The lads were also tidy, in their suits from Burton's, Hepworth's, or the Fifty-shilling Tailors. At times though, under our grandeur, lumps of mud from the football fields clung to our legs. Washing facilities were at a premium in the Dyserth League in spite of the Waterfalls!

The Plaza, Odeon, and Regal would be ready to welcome us for one and six, or one and nine. The three places looked very regal, thick carpets, plush seats, and beautiful coloured curtains in front of the screen.

Immaculate is the only word to describe the commissionaire who stood in the entrance of each of these cinemas. His costly uniform of maroon or bright blue had been adorned with yards of heavy, golden, decorative braid. He did not have much to do (any more than the chief executive in local government) except to say in his velvet-smooth voice:

'Seats at a shilling, one and six, and one and nine.'

His was a visible contribution to the oceans of splendour and luxury. He stood there, a three-dimensional advert to Macleans, Gillette, and Cherry Blossom.

This fellow had not played against Mostyn Y.M. on the Llety Field that afternoon. No, no a thousand times no. Not on your Elfed. The Llety Field of strange recollection! The one that envisaged playing football on such a field must have had Daniel Owen's imagination, and the common sense of a member of parliament. Whilst changing in the Y.M. hut, one noticed a verse which was somewhat out of place where the home team received such support,

And when the one great scorer comes
to write against your name
it matters not if you won or lost
but on how you played the game.

It mattered not indeed! In a place like Mostyn? The field was reached after climbing quite a steep hillside for about a quarter of a mile. Then total disbelief would be registered on the faces of those who saw the wonder for the first time. If they did not know the meaning of a higgledy-piggledy-pitch previously, then their education would be completed that afternoon. Here was the greatest resemblance ever to the Contour Maps I saw in the County School. It had a little of everything—slopes, hollows, furrows, bogland and where the left-back stood as his team battled towards Greenfield arose quite a hillock which we christened Hill Sixty. In truth it had everything that should not have been on a football field. The goals and corner flags were standard, the support was massive and noisy, and the players were excellent. (Here was the nursery for Mike England, Roy Vernon and many others). When I hear some of the football commentators on the television complaining about the state of some Premier League grounds, I remember, and smile.

It was pure ecstasy in the Rhyl cinemas, and the films weren't bad either. During the interval, in the ice-cream queue, we would sometimes see Islwyn Cyrnol from Trelawnyd, Alan Bickerstaff from Dyserth, Mike Bloxham from Prestatyn, or Arthur Riley from Rhyl, and we would get the results of the games they had played that afternoon. Two things were of paramount importance to us, the amount of embracing that had been enjoyed on the football field, and later in more luxurious surroundings at night.

The chocolates we ate during the evening did not prevent us from directing our footsteps towards Evans's Cafe to have a mammoth meal of fish and chips, or ham and chips and bread and butter, the butter having been lavishly plastered on with a trowel. The meal would not be complete without cream cakes (in the plural). We'd turn homewards on the top bus, and both of us would be home before eleven o'clock—the time today's courting couples start out!

After we had been courting openly for about two years, our parents saw that we were pretty serious about the matter, and I was invited to Betty's home for tea, while Betty came to Cartrefle for supper. At first these were occasional visits, but as the months and years went by they became much more regular.

Isn't it strange how every first experience sticks in the memory? That first kiss. No, not that one from Auntie Mary, but that first real one from Betty. Neither will that first goal for the 1st XI ever be erased from the memory. The recollection of my first meal at Betty's home is also very vivid, not only for the abundance of food, but for the table cloth and crockery. The whiteness of the table-cloth is stamped in my memory. Betty's mother was one of the reasons why the Robin Starch shareholders were so rich; and an occasional little professional spit on the old smoothing iron ensured that the lines of the fold of the cloth stood out like the

Tommy and Minnie Morris,
my parents-in-law.

equator. I also knew that the beautiful tea-service which was on the table that afternoon had not come from the back-kitchen. To a lad who was used to swigging water from a tin-bottle in the pit, and from N.C.B. mugs in the canteen, this experience had a rare savour about it. By now most of the beautiful tea-service has disappeared, and only a few pieces remain. On my special request Betty placed two cups and two saucers of the set in a glass cabinet we have in the parlour. The cabinet belonged to Ennis, and the beautiful contents were mainly chosen by her. Here is the beautiful Prose Medal she won in the Pantyfedwen Eisteddfod, Pontrhydfendigaid 1975 for her novel 'Y Gri Unig' (The Lonely Cry). Here is the beautiful Royal Albert tea-service she bought us. Here is the little glass poodle. Here lie memories. Here lies love. Only things which are very important in our sight are kept in this cabinet. That is why I asked Betty to include the two cups and two saucers. Betty also remembers her first visit to our cottage. It was a Sunday night—after the service of course.

It's the food she had that is foremost in Betty's memory; cold lamb, tomatoes, tomato sauce, and the raspberries from our garden, and egg custard.

'Give us a tinkle on the piano, Bet. These lads don't know the difference between the "Middle C" and the "Lost Chord". There's some good tunes in this year's *Selection of Songs* . . . smashing, Bet. What about that one on page twenty four? . . . What about that one . . .?' A tinkle on the piano was enjoyed many times afterwards. Together, supper and song.

Years later in the same cottage, the same girl 'tinkled the piano' in my father's funeral service. We sang one of his favourites.

> Now fetters are all broken
> and feet are wholly free.

When Betty played the organ or piano the singing was bound to be good, and the singing was good on that day—it was the Bass voices that were a little weak.

But back to the courting, and to one evening in February 1950. We were both in Cae Bryn, the field where I 'saw' Betty first. Although it was the shortest month, it was quite mild, mild enough for us to sit on the grass. For some reason or other we had a lover's tiff, and I'm sure that this happens to lovers of every age. Although, one overman who came to Point of Ayr from the South Wales pits swore that he had never quarelled with his wife all through their married life.

'Impossible,' said my father, 'absolutely impossible. Especially bearing in mind the wife he had. After they had settled down in Penyffordd for about twelve months Mrs James asked me if I would go there to kill a pig. I was to kill during one afternoon, and then go back the following day to cut it up, salt it, store it in the salting turnel, and all for half-a-crown. When I told her my price she told me that there was a man in South Wales slaughtering for one-and-sixpence, and the answer I gave her was for her to fetch the man from South Wales the next time she had a pig ready for slaughter. Living all your life with a madam like that without quarrelling? Impossible.'

Holy scripture has a verse which states, 'let not the sun set on your sorrows.' I doubt whether I knew that verse at the time, but that's what we did. A kiss and an embrace, and in no time the words of the quarrel were rolling over Hoylake, bereft of clouds was Cae Bryn. We decided on the spot that we would get married as soon as possible. The arrangements were made smoothly, and the wedding took place in Gwynfa Chapel, Rhewl Fawr on 25th March 1950. The bride, a mere nineteen year-old girl, becoming the wife of a twenty-three year-old coal-miner, accepting all the hard work and worry which that involved,—the early rising, the dusty, sweaty clothes, the black-leading, firing the big boiler, completing the wash without any conveniences to help her, cleaning the house, and preparing food for two who would rush to the table as if they were on piece-work. The skipful of dinner included meat, potatoes, carrots, and then a jam-roly-poly pudding and custard which would disappear before anyone could say 'cholesterol'.

Bearing in mind the importance of the football field during our courtship the caption which appeared above our wedding photograph in the Rhyl Journal was not unsuitable,

Rhyl F.C. Player Married.

Into Heights of Parenthood

It was on the Damascus Road that Paul the Apostle had the greatest experience of his life. It was on Picton Road, in Brooks' Shop telephone kiosk, that I had my never-to-be-forgotten experience.

I believe that the first day of Spring falls on 21st March, but to us in Meddfod it came in wonder and exquisite beauty a day late in 1953. With her usual thoroughness Betty had everything prepared for the arrival of our first little baby. A name had been chosen for a girl, and a name (two as a matter of fact), for a boy, but only the two of us shared the secret. I had also had a word with my father, and the Austin Seven would be available.

'At a minute's notice, my golden lad, anything for Bet.'

We went to bed at about ten-thirty on the Saturday night, as usual, and I slept soundly until I was awoken about five-thirty in the morning.

'I think the pains are starting.'

'Fine, Bet bach, you stay here, I'll go down and brew a cuppa.'

Whatever the occasion, happiness, sadness, celebration or crisis the Meddfodians must have a cup of tea.

'Two spoonfuls of sugar for you today, Bet. Are you still getting the pains?'

'No worse, but I think it would be better for you to fetch the car.'

Usually my father never worked Sundays, but of course, owing to some emergency he had to go in on this Sunday. My Father and the car in the Point of Ayr. I knew that he did not carry his keys with him down the pit, but left them with the lamp-room attendant. I jumped on my bike, and went full speed to my sister Rhiannon's house in Ffynnongroyw, to ask her to prepare herself to accompany us to the hospital. I then rushed to the pit and got the keys from the lamp-room attendant.

'Little feet making big feet run, Einion?'

'Aye. Thanks, Bob. So long.'

I then rushed back, calling for Rhiannon on our way to Meddfod. Betty has always been totally self-composed, and after beating an egg into a tumbler-full of milk she drank it carefully before we left for St Asaph. Rhiannon had been nursing in Liverpool for a while and she knew that the professional greeting to a nurse that we met in the entrance was 'For admission, nurse,' and we were obedient to the professional, 'Follow me,' we heard in reply.

'I'll phone here at two o'clock, and I'll come here tonight.'

'Fine, and Mam will be coming with you tonight, and your mother tomorrow night; and your pit clothes are in the case in the back-bedroom ready for you to go to Picton.'

No one could arrange things better than she did.

No one was braver than she was.

'Till tonight, Bet bach.'

'Till tonight, Einion.'

Rhiannon was taken back to Ffynnongroyw after we had informed Betty's parents of the situation. Then to the Point of Ayr to return the old Austin, then to Picton on my bike. It was there that I intended staying for the ten-day period, although the same welcome awaited me in Bryn Teg also; I could eat with my feet in the trough in both houses.

Neither family had a phone in the house, therefore, after dinner, I went to the telephone kiosk near Brooks' Shop by two o'clock. I possess plenty of faults like everyone else, but no one can accuse me of not being punctual. Because public property was never vandalised in those days I was able to phone the hospital with ease, and the answer I received to my first question was,

'She's just given birth to a baby daughter, six pounds nine ounces.'

The answer I got to my second question was,

'Yes, they're both well.'

That was my great moment. Both of them well. I cannot describe my feelings, but I know that that was the most exalted experience I ever had and that I shall never experience its equal again in this world.

'I've only come to the road to have some fresh air,' said Betty's mother when I arrived. But I knew that she could have the same fresh air in the back-garden if that was her only intention.

'A little girl, six pounds nine ounces and they're both well.'

'Oh very good, come in for a cuppa, Tommy's in the house, he usually goes for a nap in the afternoon, but he wasn't sleepy this afternoon . . . a little girl, Tommy.'

'Well thank goodness, I say.'

Cuppa.

'We'll see you tonight at seven.'

'Yes, the same time that I'd arrive when we were courting.'

'But Dick Barton won't be on the wireless tonight, it's Sunday.'

Mam had also felt it a little stifling when I met her at the front gate of the cottage, and my father who usually slept like a tortoise after dinner, had decided to listen to the wireless, broadening his wireless horizons, which only usually encompassed the Community Hymn Singing at tea-time. The same happiness was shared here also, and of course it was essential to have another cup of tea.

That night, after reaching the ward, we saw that Betty was comfortable. The mothers and babies were kept apart those days. The little bundles would be brought on trolleys to be fed by their mothers, and then back to their cots in a side-ward. At eight o'clock the fathers (and grannies) were allowed *outside* the door of the side-ward to see the babies through a small window. For the half-hour we chatted a little, and we were quiet every other lest we should tire the new mother. 'Yes, the grapes would be good tomorrow, and the Lucozade would be good now, just half a tumblerful, please,' and yes I was going to work that night, 'after I've seen our baby, Betty bach.'

It was through a little window in the babies' side ward that I saw Ennis for the first time. Our meaning, so harmonious, so safe in the nurse's arms. Then back to Betty for the farewell kiss, which was not savoured until we had both fully agreed that

'This is Love'.

One never-to-be-forgotten day, at the beginning of wonderful years for the Meddfod family. The years when our cup was running over, and although our world was small it would seethe with song.

'I'll be fine if I can have a doll.'

Ennis never desired much. When she was small—a doll. After growing up—health. She had many dolls . . .

A doll was one of the presents she received on Christmas Day 1955, as she approached her third birthday. We had to buy her a pram as well of course; that same pram which caused a small scratch on our sideboard-cupboard door. It was because of that scratch (and its wonderful connections) that I was unable to bring myself to sell this piece of furniture, when we were renewing the furniture in our living-room about twelve months ago. I settled on a compromise and put it in the wash-house—the utility room for the swells; it is still useful, it still speaks.

Of course, we had to take the doll with us when we went to Nain and Taid Morris' house for Christmas tea. It was not necessary for anyone to dream about a white Christmas, only enjoy it, because it had started snowing a little before dinner-time and had then continued. Not a snow storm, but a snow dream which kissed cheeks, where kisses had previously been showered. Past Gwynfa Chapel up to Cae Bach stile. Pausing there for a while before climbing over it. The scene will remain in my memory all my life; no one had crossed the field that day; we gazed on the purity, whiteness and cleanliness; something untouchable was enjoyed in the beauty. The Cae Bach hedges were a suitable frame around, and the

Ennis at 21 years of age.

three of us (and the doll) stepped into the picture. Because the snow was deeper in the field than on the road I carried Ennis on my shoulders. As the name of the field suggests, it was not a long journey, and we knew that a welcome, warmth, and food would be awaiting us.

'Come in, lads bach, and Happy Christmas.'

'Come to the fire, Ennis, to warm your hands.'

'Here's Judy, and she's a good girl like Ennis.'

'Was the turkey good?'

'Which Christmas pudding do you like best, this year's or last year's from Mary Edith?'

'What else did you get from Father Christmas?'

'A pram, books, crayons, sweets, chocolates, jig-saw, orange, nuts, and Dad had a lump of coal in his stocking.'

'Ha Ha Ha.'

'Doesn't matter what you've eaten, come to the table here. Your Mother's been preparing food since yesterday. You'd think she was expecting Abram Wood's family here, and not the Rhewl Fawr family.'

'Is there enough sherry in the trifle, tell me? More cream? Another piece of cake? Another cup of tea?'

'Is Nain Jones coming here for supper tonight?'

'Come to this big chair, Einion, these women will wash up . . . what y'uh working next week? . . . this team played well last Saturday . . . there'll be more work for them against Mostyn Y.M. next Saturday. They're a good team, and Elfed Ellis and Jack Langley are strapping lads, and someone will have to put a couple of scotches in Brinley Richards's wheels, or it'll be a run-away and a fall on Cae Bryn. And the Peny-ffordd backers will have to make sure that the stop-blocks are closed at the beginning or we'll have to get Wil Cyrtans to set brattice cloth across the goal.'

'Aye, or Jack y Bricky and Jerry to work in the goal, they like working Saturday afternoon to have double-pay.'

'But the weather'll have to be better or we'll have a game of snowballs.'

'Ask Nain for more sweets, love. Mini, have you got a drink of that egg-flip for Einion and Betty? I know the sherry's all gone in the trifle.'

'Good health.'

'Betty's by the organ on Sunday isn't she, and the old Jones-boy preaching. D'y'uh remember him knocking on this back door, and you stripped naked in the little bath in the back-kitchen?'

'Yes, very well, and I remember what I shouted 'cos I didn't know that it was him there, I thought that Betty was pulling mi leg. But that's it, he's got to have one like ev'ryone else.'

'Hello Nain Jones, how are you?'

'Tidy thank you. Are you pretty fair?'

Somehow or other we succeeded to eat the supper as well. The snow continued to fall, and when Taid Morris returned to the house after locking the coal-house he said that it wasn't fit for us to think of going home that night.

'Hooray, sleeping in Nain Morris's place.'

That's what happened, Ennis and Judy by the wall, Betty in the middle, while I became a sort of safety-sprag on the edge.

Truly a white Christmas.

A snap of Cae Bach in snow . . . Cae Bach and the green, green grass. The second picture has the same frame, but winter has been changed for summer; a student of the University College of Wales, Bangor, replaces the little girl; Scilti, the white poodle replaces Judy. Ennis throwing a ball, and Scilti running after it to retrieve it. Throwing the ball again, and Scilti disappearing in the waves of the lush grass before finding the ball; before returning, ball in mouth, and dropping it by the feet he loved. Great pleasure, laughter, Love. The playing ceased in about half an hour. From that day, while Scilti was able to walk, he went for a walk up to Cae Bach stile, but did not wish to go further, he would only raise his two front paws on the stile, look over it for a few seconds and then turn back for home. Scilti also had a good memory . . .

I think a lot about Cae Bach. I think a lot about Heaven. As far as I can see there is in Heaven, as in Cae Bach, purity, whiteness, beauty and green pastures, but, most important of all, our beloved Ennis is also there, and little Scilti.

An Avowed Eisteddfodic

'Let the Bard be seated in the Peace of the Eisteddfod.'

Those words have been woven in my mind since childhood and I have had a burning desire to win the National Eisteddfod Chair since I was about four years old. I know when the seed was sown. I also know about the watering and the labour, and by now the 'Eisteddfod Harvest'.

While I lived in the cottage in Picton there was no water indoors, apart from the drips which fell through the back-kitchen roof at times when it would be raining cats and dogs. The seven of us used a small galvanised bath in front of the living-room fire in winter, and in the back-kitchen in summer. During our childhood, Saturday night was the great-scouring night, in readiness for chapel on Sunday.

Dafydd Jones's family, who lived in the farm nearby had purchased a bath twice the size of ours,—a bungalow bath. This was kept in the wash-house on the yard near the big old-fashioned coal boiler. Occasionally Mrs Jones would shout on Mam to let the children go there to have a bath in the big bath. This meant that about seven or eight of us children could be scrubbed in one afternoon, something similar to sheep-dipping time. Because two children could be 'dipped' together, the time would be halved.

After being partly dried, we were allowed to go into the farm-house to complete the process and to await the rest of the family. To complete the treat, we were allowed to listen to the gramaphone, the only one in Picton at the time. The farm family was renowned for pioneering; they were the first in the locality to have a camera, wireless, tractor, gramaphone, a motor-car, and later, a television.

Although they had a motor car it was never driven by Dafydd Jones or his family. It was kept in a hut on the yard, and occasionally a man named Mr Parrish from Rhyl would travel the eight-mile journey on the bus to Tan-lan, walk up the Brêc, and then take some of the family for a little ride in the car. The car would then be returned to its hut, where it would remain until Mr Parrish's next visit. Someone had a bargain when they bought this car. In this case, the words, 'one owner, low mileage' were not a lie.

Gracie Fields and her massive aspidistra featured on one of the records. Richard Tauber and Caruso were also there amongst the scores of other records, many of them bearing the appealing label of the dog listening to its master's voice coming through the loud-speaker. Contrary to expectation, in the middle of this musical feast, there was one record of the Chairing of the Bard ceremony in the National Eisteddfod. I have no idea who was chaired, but I know that that record left an indelible impression on a four-year-old lad. My, oh my, this was the one for me. Besides its most pleasant appeal,

Picton Yard Farm, where I heard the record of the Chairing Ceremony. David Wright, one of Dafydd Jones's grandsons in the photograph.

the attraction of Gracie, Tauber and Caruso would 'become subdued, and silent be'. I had no desire to be a singer, but from that day when I heard the record of the Chairing Ceremony in the farm, I knew which goal I would have to strive for. The cottage and the locality were immersed in singing; that was also the attraction for the family when we went to the Chair Eisteddfod in Ffynnongroyw; but to me the great thrill was to see the Bard rising to his feet in the rear of the Chapel. I would stare at him, and think how wonderful it would be for me to be in his shoes. I never desired to wear Hugh Lloyd, the tenor's shoes, nor the bass-singer's Rowland Bowen, only the shoes worn by that unfamiliar, strange, dreamy man who was in the rear seats of Bethania.

The seed was sown early, but the shoots did not appear for years. I always enjoyed the hymn-book. Reading and learning poetry was a pleasure in school when that would not come into conflict with the footballing. When I was about eighteen years old I remember writing a few verses to tease my mate, Don Brooks,

> The festive mood in Gwynfa will be grand
> with Donald and Eluned hand in hand.
> Bill Davies will be cat'ring,
> we'll all be titivating.
> The festive mood in Gwynfa will be grand.

I know that I wasn't a poet at the time but at least I prophesied correctly.

When I was twenty-two years old a Competitive Meeting was held in our Chapel. This was something of a little higher standard than the Penny Readings, and slightly lower than the Chair Eisteddfod held in Ffynnongroyw on St David's Day. Usually, the sum total of the literature section in the competitive meeting was to complete a limerick, but this year, the committee had been extremely ambitious, and added a competition for writing a lyrical poem on the Garth Woods. The winner was offered a silver cup, kindly donated by Mr and Mrs Gwilym Lloyd. I had a desire to compete, maybe because I knew more about the subject (one of the diverse lovers' lanes) than I knew about lyrical poems. The adjudicator was the Rev. R.R. Jones, Prestatyn—I knew nothing about him either. Not that it mattered, but I knew more about the music adjudicator, Mrs Gertrude Wilson Ellis, Gwespyr. Here was an adjudicator who never had a doubt in her mind, because she would finish her adjudication every time with the positive words,

'Without ifs or buts I award the prize to . . .'

Strangely the prize for the poem was awarded to me, who had hidden under the nom-de-plume Son of the Cottage. The feat was accomplished in a group of five competitors—to call us poets would have been utter nonsense. It was a great experience, to walk from the rear of the Chapel to receive the silver cup. I knew even in those early days that every proper bard sat in the rear seats, in order to savour the long journey to the platform. The cup was small, and it is not with any disrespect that I say that it was not spanking new. To me, there is something of the Holy Grail mystique about it, casting a silver gleam, a gleam that has its source in somewhere better than the 'Silvo' tin. The cup is still one of my treasures.

I wrote a poem about love, but I had to follow the fashion, and bring my sweetheart's death into it. A poem about Love full of Death. Years later I wrote a poem about Death which was full of Love.

The 'devil of song' was not unknown in our locality. Sometimes the reserve precentor had forgotten more about music than the precentor had ever learnt; and the 'name-that-tune' organist demanded to play during the big-night of the Preaching Meeting, and sometimes the choirs would have a raw deal because the adjudicator was drunk. Something similar came to the 'poets world' the night I won the cup. I walked in the dark down Rhewl Fawr behind two old men who were 'in' the competition. They were not aware that I was a few yards behind them, although they knew I was an inch or two ahead of them in the competition.

'Einion won the cup tonight for the poem.'

'At least it was him who went down for the prize.'

I do not blame them in the least. I know that it was difficult for them to believe that one that had used his feet so much was now starting to use his head and his heart.

For some reason I did not compose further for about four years, until Ennis was born. In the middle of our bliss the poems began to flow. A poem to Ennis to start with entitled 'Breaking a Promise', and no Shakespeare has known pride to surpass mine, when I saw the work in print in a denominational magazine, and the Rev. Gwilym R. Tilsley was generous and kind in his praise. A busy period of writing. A period of buying poetry books from one little shelf in Lyn's Antique Shop, Rhyl. A period of competing for a chair; the thrill of winning a cup was not quite the same as winning a chair. My first four attempts were unsuccessful, but suddenly on 2nd January 1956 the dawn broke, late at night.

A Chair Eisteddfod was held annually on New Year's Day in a village called Rhydlewis. Before this time I had never heard of the place. The Cilie farm was not in my geographical knowledge, and the names Jac Alun, Isfoel, and T. Llew Jones were not part of my general knowledge, and I'm sure that Dic and Tydfor were but schoolboys. At that time Alan Lloyd Roberts knew more about conkers than about 'cynghanedd'. The competition was for three lyrical poems entitled 'Dreams'. If I was not familiar with the Rev. R.R. Jones, Prestatyn, who awarded me the prize in Gwynfa, my prospects of knowing anything about the adjudicator in Rhydlewis were far more remote. He came from somewhere far more outlandish. He came from some other continent called Allt-y-blaca! I forwarded my entry, and waited for the post. By now, through years of experience, I can differentiate between the Royal Mail Van and a hundred other vehicles that pass the front door. Because New Year's Day fell on a Sunday that year the eisteddfod was held on the Monday. The postman failed to call on the Saturday, and I started to fear that all my dreams about realising the thrill of the H.M.V. record years ago were in vain.

A Watchnight was held in Peniel Schoolroom that Saturday, and Betty attended. I stayed home to look after Ennis, who was sound asleep by this time. I was enjoying reading a book of poems entitled *Cerddi'r Encil* by Cenech. I was reading my favourite poem of that volume, Birkenhead's Black Chair.

Now you who relish warfare today high Heaven is sad
What was the gain to Britain of losing this young lad?

My memory stretches like elastic regarding events that occurred forty years ago, and breaks like frayed cotton regarding things that happened forty minutes ago. About eleven o'clock I heard heavy footsteps approaching our back door, and an equally heavy knock on the door at once. It was quite safe in those days to open the door. It was Bob Roberts, Rhewl Fawr Farm, and after coming in he told me that he had received a message from the Ffynnongroyw policeman. Because the policeman was English he had not understood the message properly, but somebody from South Wales wanted me to go down to fetch a chair, or something. Bob-y-Rob failed to see much sense in the message, either. It made more than sense to me.

'Thank you very much Mr Roberts. Good night Mr Roberts.'

The wonderful moment had arrived, and no one in the house to share in the ecstasy of the experience. It would be a shame to wake Ennis, and my better-half was in Peniel Schoolroom. There was nothing to do but act as I had done many times in a crisis, and knock the wall on Auntie Ann next door. She arrived post-haste.

'What's the matter?'

'There's nothing the matter, Auntie Ann bach, but I've won a chair in South Wales.'

'In South Wales? Where's Betty?'

'In the schoolroom.'

'Wait a minute and I'll go down and tell her!'

The watchnight proceedings were interrupted, and Einion was given three cheers on his accomplishment there and then. During my schooldays I read a poem entitled 'How they Brought the Good News from Aix to Ghent', recalling the exploits of a soldier bearing the good news on horseback. Auntie Ann delivered the good news in her slippers. We are not related. It was through her kindness and good deeds that she became, and remains, our auntie.

The following day a telegram arrived saying,

'Congratulations, Chaired Bard of Rhydlewis.'

Ennis was not three years old at the time but she understood that something rather special had occurred. About a fortnight previously my sister, Leah, had broken her leg, and my parents had gone to Liverpool to give a helping hand with the five children. Because Leah did not possess a phone I left a phone message for her with one of her neighbours, informing her of the good news. About five o'clock, when the Meddfodians were having tea a sedate little knock was heard on the front door. It was Mam, and in the road, in the back seat of a taxi sat my father, as proud as the Aga Khan. Prior to this day they had not been in a taxi all their lives, but on this occasion the taxi clock was ticking all the way from Liverpool.

'Leah'll have to put salt on the work for a while because we're going with Ein to South Wales,' was his ruling. The only one who could ever change his mind had been buried a few years previously.

It was in a taxi that we went to Rhydlewis too. A brimful taxi, Ennis, Betty and myself, Betty's mother, and Auntie Mary Edith, Mam and my father, and the driver of course! It was a long journey to Cardiganshire, and although the chairing ceremony had been arranged for nine o'clock it was nearer midnight before the ceremony commenced.

The adjudicator, the Rev. D. Jacob Davies, said that he sensed that there were better poets in the competition than 'Coliar', but somehow he was compelled to come back to his collection every time; and I remember the last line of his adjudication, '. . . and although it is a posy of heather flowers he possesses, they're pretty, they're alive, they're growing.'

Yes, 'Coliar' was called upon to stand up, and no one was ever more ready to obey. I stood up in the rear seats in the Y.M.C.A. Hall, exactly the same as those poets in the Ffynnongroyw Eisteddfodau years ago. It was hard to understand the locals speaking, but I understood the adjudicator and the master of cemeronies with ease!

We arrived home at five o'clock in the morning, to see my mates waiting for the bus to convey them to the pit.

'See yuh boys! Yu'll have to find someone else to blast the coal for Elwyn Luke today.' Although Ennis was not three years old at the time, she remembered the occasion vividly all through her life.

The Rev. Dafydd Owen, Mold guided me through the intricacies of the strict metres, I took part in poetic trials (*Ymryson y Beirdd*), and I continued to compete. Ennis and Betty have been a constant support to me through the years. I remember one National adjudicator saying something pretty cruel about my attempt.

'Don't listen to him, Dad,' were the words of consolation I received. Stubbornly, I refused to listen to him. It was also a stronger and more appealing sound that I could discern in the voice of the H.M.V. than the H.M.I. We did not fill a taxi for every Eisteddfod, but there were three of us together in most of them.

Apart from supporting her father, Ennis also enjoyed creating literature. Her influence remains. She won many prizes for short stories, essays and a novel. It was for her first novel *The Lonely Cry (Y Gri Unig)* that she won the Prose Medal and £150 prize money in the Pontrhydfendigaid Eisteddfod, while she was a student at University College of Wales, Bangor, competing against a number of 'old cobs' according to what we heard afterwards. For a short-story entitled 'Sally's Friend' (*Ffrind Sali*) she won the first prize outright in the National Eisteddfod at Dyffryn Lliw in 1980. Forty-nine

Eisteddfodic trophies, including the Pontrhydfendigaid Eisteddfod Medal won by Ennis.

competitors and 'Without a fire, Without a bed' winning. Around this time it was not unusual for an adjudicator to share the prize between two or three in this competition. It is with respect and pride that I say that the prize was not shared in 1980. The prize of £50 was offered, and that's the amount that came to Meddfod, and although Ennis, because of her illness, was not earning a wage at that time, she insisted on sharing the prize equally—yes, equally, with the yellow-skinned children who live 'Far, far in China and Isles of Japan'. Twenty-five pounds of the prize money went to a hospital in China. We received the address from the Rev. John Tudor, who officiated in Holywell at the time.

I came close to the top for the National Eisteddfod Chair a few times; indeed I had one hand on the arm in 1971 in Bangor, according to one of the adjudicators, but as every poet knows the important thing is to get another part of one's anatomy on the Chair.

I entered an Ode to the Wrexham National Eisteddfod, and thought I stood a good chance (but for that matter, isn't that what every poet thinks every year?)

'Don't you think it would be better for you to have a new suit, in case?'

'I'll buy Dad a new suit.'

With her usual kindness that is what Ennis did, she bought the dearest suit I've ever had, and as a bonus she bought the shirt, tie, the socks, and the shoes. Yes indeed, the whole lot.

During my days in the County School one of my favourite songs in the music lesson was 'Passing By'. That is what happened to the Post-Office Van that year also. It was Donald Evans who wore his best suit in Wrexham. I did not wear the clothes I had from Ennis, but kept them carefully. I did not compete for the National Chair afterwards either until 1983 when the Eisteddfod was held in Anglesey. It was there that I was honoured to wear the beautiful clothes that I had received from our beloved Ennis—only the tie was different.

More About my Heroes

They've now all gone from the Point of Ayr,
void is the commune that once was there . . .

Jonathan Edwards, and Will Lei Joss,
Will Parry Asaph, Will Parry Moss.
William Jones, Bagillt, and William Glên,
William yr Injian, and Twm Llai Mên.
Gwilym y Caeau, and old Ned Chwain,
Reuben y Ffitar and Tommy Nain.
Jac y Bricci, Jac Blythin y Go,
Barnet y Sadler and Sei'r Iard Lo.
Ifan yr Ostler and Dafydd Dew,
Jac Twm Twmi and Ned Parry Llew.
Ifans y Gaffer, Jac Wil Trai Mi,
Jo'r Hen Stabal, Foxes and John B.
Sam Hughes y Teilwr and Dan y Nant.
Tommy Epworth, Will Lloyd y Pant.
M'redydd y Moelfryn and Will Strêt Back,
Old Tommy Lady and Big Tall Jack.
Jack Garreg Lwyd and John Bo Bo,
Ned Parry Scinar, Trefor Pen Rho.
Daniel Lei Burum, many a Gec,
Emlyn y Lamprwm, and Dic Coes Glec.
Will Lloyd Eco and Dafydd y Coed.
Harri'r Mochyn, and Pitar Wyth Oed.
Alf Iard Butyn and Gwilym y Groes,
the old Ned Wedyn and Hywel Toes.
Will Black Padan and Ned Twll Tân,
Tommy Ffordd Ddŵr and Tommy Crys Glân.
The Old Ifan Beti, Jo Glan Llyn,
Tomi'r Goler and Ned Trowsus Gwyn.
Bob y Drymar, Dan Green and Ned El,
Old Ned y Bocsiwr and Ned y Swel.

Point of Ayr miners buying their poppies in 1935. It must have been a Friday, pay-day. Before the days of the canteen the miners never carried money to work. Neither did they carry much money from there.

Bob Tan-y-Capel, Bob Pinci Ponc,
Robin Daily Mail and Wili'r Bonc.
Bob Hughes from Trefor, Will Ben Relwê,
John Williams Point, Will Griffiths y Te,
The old Jo Twmpath, Clifford Da Da,
Huw Williams yr Huts, Jac Bacyn Da.
Twm Parry Cyrnol, Isfryn y Gors,
Daniel Jones Mwga, and Pirs White Hors.
Ned y Leitaws, Edward Jones y Dre,
little Jac Ffantêl and Twm Ipidê.
Void is the commune that once was there
they've now all gone from the Point of Ayr.

Names, and men all familiar to me, each Jack-one of them, and I worked with most of them. By now not only have they left the Point of Ayr, but they have also gone to a 'better world to live'. Only two or three of them remain.

This poem appeared in *Y Cymro* newspaper a few years ago, and in no time I received a letter from a research student from Brittany, who had come over to Wales to further his studies, asking permission to include this in his thesis. I agreed happily to his request, and it's now great to think that these names are now together, my forefathers, my family, my mates, in an University in Brittany. None of them ever received a University Education, but I agree with what Wynn Spenser, previously Headmaster of Ysgol Penmorfa, Prestatyn told me a few years ago in conversation about the old miners. 'Some of them had been gifted with talents suitable for the University, the only thing missing was the opportunity.'

When Betty and I go on our frequent visits to Picton Cemetery I often think along the same lines as the English Poet when I see some of their names (not their nick-names!) on the grave stones.

> What mute ignoble Milton here may lie
> what Cromwell guiltless of his people's blood.

When I read the poem on occasion a fall of recollections come to mind. The origin of the nick-names is interesting. I recall hearing Emyr Humphreys once saying in a lecture that Dylan Thomas even had never devised a better name than Pinci Ponc. When Robert William Lloyd brought two kittens to the house, he called one Pinci and the other Ponc. The cats and their owner have gone since years, but the nick-name remains. I am thrilled by the name Jac-Will-Trai-Mi, although that of his son, Cyril-Trai-Mi does not appeal so strongly. I have no idea how this name came to be.

More's the pity, I know too well how the name Will Griffiths y Te came about. I am saddened every time I hear the name. After the 1921 and 1926 strikes the miners were not allowed to re-start work together. It was necessary for them to come down daily to the office-door, knock, remove the cap, and ask for work. Sometimes two or three of them would be given permission to re-start work, and they would receive an authorisation paper for a lamp, but the cruel answer blurted out by the manager to the others was 'No more, come back tomorrow.'

Will Griffiths had been prominent in the union—too prominent in the sight of the bosses, and therefore there was many a 'tomorrow' in his case, tomorrow . . . tomorrow.

Many pairs of his boots became nothing but uppers, and in order to obtain a few shillings to support his wife and daughter Blodwen he would visit his friends' homes with a caseful of Kardomah Tea to sell. A cup of tea and a chat here and there. I remember him having many a tea with us in Picton long ago, nothing special, home-made jam butties, and an oven-bottom cake, or a slice of bread-pudding. Years of unemployment can make even a crust seem tasty. He was a gentle, kindly soul with a tenor voice sweeter than sugar. His body was scarred in the pit, but not his soul.

The nick-names did not worry some of their bearers, but some hated the nick-names thrust upon them. The Foxes hated their tails, especially if the name was shouted to the accompaniment of the fox and hounds' horn, and there were some pretty goood impersonators in the pit. On the other hand Will Cannon did not give the impression that carrying his warring weight worried him. Often, when a wag would come towards Will, the performance would be for the oncomer to shout,

'There were cannons to the left of them,' and Will's answer, without fail, would be, 'And cannons to the right of them, by the heck.'

Good old Will, he was a deacon in the Wesleyan Chapel in Gronant, and consequently abstained from using the stronger version of the last word.

It's recorded in the Bible that God visits 'the fathers and the sons until the third and fourth generation'. Some of the Point of Ayr nick-names have the same duration. A while back I saw Bob Wocyn in Prestatyn, and while we were chatting, I asked him about the source of his diploma.

'Oh,' answered Bob, 'easy, my grandfather on my father's side lived in Mostyn in a place called Ladies Walks, the name "walks" became "walking", and "walking" became "Wocyn".'

Bob is a little over his three-score-ten-years,. therefore his grandfather's time goes back well beyond a hundred years.

Ned Baco Main's family is in the same category. Edward Jones was the first to carry the name because he went ill after swallowing his father's chewing baccy. A while back I had a very pleasant experience with one of the descendants of this dear and well respected family. Our coal bunker was finished, and I thought, being as so many people were using gas and electricity to heat their houses, that I might be able to get a second-hand bargain. I searched through the 'for sale' columns in the local paper, and saw one being offered for sale in Prestatyn. I picked up the phone. A lady answered, in English of course. The answer was affirmative. Her next question to me was,

'Where do you come from?'

'From Penyffordd.'

'Oh, I'm from Penyffordd too. I'm Linda. I'm Twm Baco Main's daughter.'

Good for you, Linda, for being proud of your roots. Living as she does in an area falsely exalted by Mrs Davies the Laurels, Mrs Jones-Morris the Cedars, Mrs Roberts the Bank, she still remembers the rock she was hewn from. The bunker became mine too!

The name M'redydd y Moelfryn springs to mind because of some of his sayings. His nick-name has no special significance apart from recording his birthplace. After becoming married he went to live in a small-holding called Pen-draw Farm, on the Glasdir, and he also worked in the pit. His duties in the pit were those of a fireman, and I have often heard him, as he called the loaders back to their work after eating their snappin, 'Come on boys, think of your living.'

'Have yuh fired that bench well for us M'redydd Williams?'

'Yes indeed, lads bach, there's an aunt in the coal for you today, yes indeed, and there's enough for John, Jane, and Charlotte. Come on boys, think of your living.'

He was the only one I ever heard using the phrase 'there's an aunt in the coal', which meant that there was an abundance of coal available in convenient lumps to load into the tubs (drams).

Alf Tai Trap, my cousin, Dan Burum, Donald Relwê and myself were pretty big mates during the early years in the pit. We were mischievous, yes, but no more. Alf spun a good yarn, Dan Burum was the local Wil Bryan, only no Daniel Owen has recorded his antics. Donald was a happy, stocky lad, with a smile on his face as wide as a passbye. During our friendship the repair shift was worked in the afternoon, and the hooking shifts in the morning and nights. For some silly reason we would come down in the last cage every time. As we descended the Two Yard Sling, and came close to the lampstation, where men gathered before being dispersed to their different work-place, I often heard one of the old miner's remark,

'Here they come, the hooter's handle, pon my soul.' This term was used to describe a person who was late for work.

During this period we were pony-drivers and as we entered the stables to collect our ponies, it would be our pleasure (when Dafydd Jones was the ostler) to turn into a musical quartet, and set new words to the tune, 'She'll be coming down the mountain'.

Which pony will you give us, Dafydd Jones?
Which pony will you give us, Dafydd Jones?
Oh give us one that's handy,
that's strong, and is not bandy.
Which pony will you give us Dafydd Jones?

The words were composed by my cousin, Alf. He had a slight grasp of the rudiments of poetry, but never bothered to improve his talent. Our rendering was not so pleasing to the musical ears of Dafydd Jones.

'Go to your work, you terrors, before I put this cane-brush on your backs.'

It is to the organ in Ffynnongroyw Baptist Chapel that we must give thanks that he did not call us by a stronger name. May his dust be in peace, and his soul in Heavenly Peace. If we were rather slow arriving at our workplace, once we reached the coal-face and took our shirts off no one could say that any of us had an idle bone in his body. We worked like slaves through the shift.

I'm sure that we were the last to 'put finger in cap'. This is an old custom, but I now doubt whether any of the present generation of miners know the meaning of the phrase. It was not an easy matter to descend into the bowels of the earth at any time, but on beautifully fine weather it was much more difficult. At times the magnet of the unspent July sun on the Brêc and Glasdir paths would be far stronger than the small oil-lamps in the far end of the Durbog, or Two-Yards; and being with a pony cannot compare to being with a sweetheart. Discussion.

'What about going home?'

'O.K. Finger in cap.'

One of the number would then throw his cap down on the ground (or later the canteen table) and everyone that was game to go home would place his index-finger in the cap. The idea was to share the responsibility, something similar to the 'round robin' at the end of a letter.

'Sweethearts, here we come.'

Recording one incident about Dan Burum must suffice. During the great snow blizzards of 1947 it was compulsory to struggle through the elements to work. It was necessary to attend work or lose pay. By the fifties, after the introduction of buses to convey us to work, the Union succeeded in obtaining an agreement whereby we would be paid if the buses did not arrive. During one of these winters heavy snow fell again. The buses could not travel up Llinegr Hill, and for a while we were paid for doing nothing, exactly the same as some N.A.L.G.O. workers! This was too good to last for long, and the manager came up with the idea of conveying us to work in army lorries from Kinmel. While the snow lasted that was our mode of travelling. It was not as pleasant as staying home, but at least it was better than walking. There was no difficulty whatsoever in becoming friends with the soldiers, they were also men under authority, men who had faced dangers, men who had been used as cheap pawns to enrich their masters. At the end of each week a collection was made for the driver, therefore every soldier was a volunteer on this call of duty.

One morning, as we returned home from the night shift, after reaching the top of Tan Lan Hill, and before reaching Gelli Bant Crossroads, Daniel spotted a pheasant in one of the fields. No one else saw the coloured delicacy, but snares, ferrets, and guns had been our pal's toys since childhood. That night he brought a double-barrelled gun and two cartridges with him to work, leaving them in the lamproom, and picking them up in the morning. The driver received precise instructions; at no cost was he to stop at the top of the hill, only drive a little slower than usual. Everything went according to plan and the ace-sniper took aim from the rear of the lorry. The second barrel was superfluous to requirements, and it could be said of the pheasant, as was often said about the aircraft of the period, 'one of their number failed to return to base'.

In the snappin-place that night a beautiful aroma escaped from one snappin-tin. No one exclaimed 'Ah, Bisto' in the underground Savoy, but that night many a one exclaimed 'Ah, pheasant'. Daniel had manoeuvred everybody to sit on his inbye side, so that the rich aroma could be wafted on what little air was available. At the end of the feast, and before giving the horse the crusts, the crackshot placed the lid on his tin with a satisfied click; and with pride in his voice he rephrased a proverb he had learnt in Picton School.

'What's good for Lord Mostyn is good for Dan Burum.'

Alf died a few years ago. For a while Dan emigrated to Australia, but failed to stay there. According to Dan he returned because of his great longing for the Old Country. Not enough rabbits for him there according to one local wag. Donald still smiles, especially when we re-live the days of innocent mischief.

At various times many seams were worked in the pit—Y Cant, Bychtyn, Two Yards, Three Yards, Five Yards, Stone-coal, Durbog, New Durbog, Old Durbog and the Five Quarters, each having a special characteristic. All seams have a good headroom (height), the lowest being the Five Quarter Seam with its headroom of five-foot-six at the coal-face, and the highest the Durbog which was at least nine-foot. The very low seams were not worked by us as was necessary for Wilf Two-foot's grandfather many years ago in the Durham pits. A very thin layer of stone ran through the middle of the stone-coal seam. There was a good roof in the Durbog and Two-Yards, but a suspect roof in the Five-Quarter and the Stone-Coal. Because the condition of the roof could change quite quickly, it was necessary for the miner to test the roof often. This was done through striking the roof with a blunt part of his pick. When the roof was good the resulting sound was a pleasant ring, and it was said that the roof was singing. When the testing resulted in a heavy, dull sound it was said that the roof was like a pudding, and sometimes when things went to the extreme it was like a flaming rice pudding, and more props had to be erected urgently before the place became smithereens. These props were our safety in the depths of the earth, and because we always took a small piece of a prop home with us each day in the depths of our poacher's pocket to kindle the fire, they also helped to keep us warm.

Coal-mining and Singing Songs

Sometimes the roof would be singing, but at other times the roof would be raised when the miners congregated to sing, in a chapel, eisteddfod, or pub, or very often in the pit itself. There's no two ways about it that the Point of Ayr area was immersed in music for many years. It is a pity that I am compelled to use the past tense in that last sentence, but the truth is, of course, that things are very different by now. About forty years ago it became fashionable for young couples in Wales to leave the villages and live in the towns. For my own part, a working-class creature, brought up on scouse and sheep's head broth, I would find it very difficult to settle down in any smoked-salmon area. Because of this exodus, and because the people that replaced them were different, this region, like many other regions, suffered a decline. A bookie's shop has replaced the butcher's, and en-suite flats have replaced the Baptism Well in the Baptist Chapel. The only conversions at present are chapel conversions into dwellings, and not the conversion of members for the indwelling Spirit. By now hardly a handful are 'praying for saving of souls'. By now the call of the bingo is more effective than the Call of the Gospel.

But forty years ago and more there was singing here, yes siree. Ffynnongroyw boasted of three good choirs, the children's choir conducted by Robert William Lloyd, the mixed-voice choir, conducted by Bessie Williams Parry, and the male-voice choir, conducted by Thomas Edward Jones. The villagers proudly boasted of three pipe-organs, and it was never a competition of 'Name that tune' when Thomas Jones or Pit Elder were seated at the organ. I very much doubt whether these able musicians ever received any money for their work, if they did it would be a trifling. Within a mile lived the uncomparable David Lloyd, who enthralled countless thousands with his wonderful voice.

Ffynnongroyw Male Voice Choir with their conductor T.E. Jones and their accompanist Miss Gwladys Hughes.

Bryn Siriol, David Lloyd's birthplace.

Some of the present residents of Ffynnongroyw decided to show their appreciation of R.W. Lloyd's sterling service, by presenting an annual prize in his memory in the Urdd National Eisteddfod. These are the types of persons worth honouring, the type that should be conspicuous on any map depicting the National Eisteddfod area. I believe that this annual Eisteddfod Map should reflect the culture of the area, and not create gimmicks. 'He that hath ears to listen, let him hearken'.

The three choirs competed very successfully in Eisteddfodau, and it was with pride that membership was attained. The old miners never tired of recalling and repeating some of the choir's feats. One story heard often, and that from different people, was one about the male-voice choir. A test-concert was held in Treuddyn, and our local choir decided to compete, eighty choir-members being conveyed in two charabancs. The festival was held in a marquee which had many entrances to it. Unfortunately, the choir arrived a little late, after their competition had commenced. Indeed, five of the choirs had already sung, before the arrival of the sixth choir. The Festival Compere called the Ffynnongroyw Choir, but there was no response. 'For the last time, is the Fynnongroyw Male Voice Choir present.' By now they had rushed in through the numerous entrances, and they all shouted 'We're here, We're here,' from all corners of the marquee. They took the platform, each one in his black bow-tie, and their usual suits. They could not boast of owning blazers—only of wonderful voices; and sang the test-piece, 'The Martyrs of The Arena'. This is the piece that refers to 'the roaring lions and hyenas'.

The adjudicator was lavish in his praise for the very high standard of the competition, and especially for the thrilling rendering by one of the choirs, and as he awarded them the prize-money and the Silver Cup, he uttered one phrase that pleased them immensely, and which remains until today music to the ear to those of us who appreciate our roots. 'This choir came in like lions, and they sang like lions.'

After Thomas Jones died the choir was silenced for a few years, until it was re-established by Edward Ellis (Ned El) from Glan'rafon. Ned El worked in the pit. He lacked the genius of his predecessor, although he had conducted the 'Gwespyr Silver Band' for a while, and had a song in his heart. Some initial success was scored, although at the moment I cannot recall in which Eisteddfod this happened, although I was a member by this time. Nevertheless I well remember an incident in work the following Monday. The selected piece was, 'Crossing the Plain', a work that had been sung a hundred times by the 'Old Choir' under T.E. Jones's baton. A group of us sat at the bottom of the Two-Yard Sling at the beginning of the shift. In our midst sat Robert William Lloyd, who believed in his heart that it was he who should now be conducting the choir. The situation that morning was tailor-made for Alf Tai Trap.

'The Choir did well on Saturday, Robat Lloyd, didn't they?'

'Huh! Warmed-up scouse I'll be damned.'

A little success was tasted, but alas also, the incidents of the Mara Waters, and that, after we crossed the Mersey waters. It was during Lewis' Eisteddfod, which had by then moved its tent from the shop to the Central Hall. The choir had also changed its name to the Point of Ayr Male Voice Choir by this time. Lei Top was the choir's president, and in the middle of each practice he would stand up to say a few words, always commencing with the words,

'It's doubtless, Brethren . . .'

Had he not already possessed a nickname he would definitely have been christened Doubtless Lei.

The test piece was:

> When I am dead, my dearest
> sing no sad songs for me.

A new piece for us and our dear conductor, and the newness was 'of bitter taste'. There was no success whatsoever. The congratulatory remarks which the Old Choir received from the adjudicator in Treuddyn are still remembered. Alas, also remembered is the opening remark delivered by Sir Hugh Robertson in Liverpool about our effort. 'This choir of miners dug a pit for themselves, and couldn't get out of it.' Elaboration is useless. After that we found nothing in that promising oasis which had a sweet taste.

Poor old Ned El also had a strange taste on his chips that night. Literally, his days as conductor were numbered. My memory fails me whether he resigned, or whether he was given the 'clog-sffortze' somewhere in the region of his bottom doh, but in about a week Alban Jenkins from Rhyl conducted the choir, and his wife Gwyneth was the accompanist. Rhys Jones also conducted the choir for a while. While Alban and Rhys wielded the baton, the musical lamp burnt brightly, but the methane of indifference seeping through these latter years has extinguished the lamp. In those days in the pit a song would alight through the light of the lamp. The lads would congregate and the singing would never be better than in pit-bottom. Another name for the pit-bottom was *Y Llygad* (the eye) because if one peered up the shaft a very small light, the size of your eye, could be seen in the top. Everyone waited here anxiously for his turn to ride in the cage, in batches of fifteen or so, to the surface. Pit-bottom was a cold, drafty spot, and everyone huddled nearer each other for shelter. Today the weather-men on the television speak of the 'wind-chill element'. We knew about this element in the pit, although we had never heard of it by the posh name; in our vocabulary it was 'pneumonia wind'. That is why pit-bottom was nick-named 'Pneumonia Junction'.

Nevertheless the singing had warmth. About fifty in the choir, no conductor, no musical instrument, but I know the conductor of the Glasgow Orpheus Choir would have something better to say to 'this choir of miners' if he heard them singing some of the old favourites in their own terrain. Of course it was the shorter pieces that were sung, and lo, hymns were our delight. I believe the favourite was,

> We give thee thanks the true Almighty God
> now for the sacred Gospel.
> Alleluya, Amen.

The words of the second verse were most fitting to prisoners of the cruel capitalistic system,

> When captive in a prison dark and black
> you lit our darkness brightly.
> Alleluya, Amen.

Because they were aware of darkness in many dimensions in their lives, their pleas for the light of dawn in the last verse were not false.

Break forth, break forth, the ever-widening dawn,
light up each country wholly.
Alleluya, Amen.

With Christmas approaching, the sound of carols would fill pit-bottom, and would waft upwards in the cage as it was wound up to the surface. It was a unique experience to stand on the pit-head-plates those mornings and hear the sound of the approaching singing. Far-off chords to begin with, but the chords would become more audible as the cage came nearer, nearer, nearer. There was no need to be very sensitive or very religious to realise the similarity of what was happening at the Point of Ayr to what happened in the fields of Bethlehem two thousand years ago. The dust-covered miners, with their thoughts on the cup of tea and Woodbine in the canteen, unconsciously re-creating anew that old Silent Night. Silent night, Stille Nachte. Yes, there were Germans amongst us, prisoners of war, excellent workers, each of us praising God in his own language. Very like the early Christians in the Book of Acts, long ago.

In our midst were two soloists who had triumphed in the National Eisteddfod. Huw Lloyd, Gronant, won on the tenor solo, and I well remember being honoured to congratulate him and shake his hand heartily on Pit-y-Pant's Landing, down pit. It takes much more than child's play for anyone to rise from Pit-y-Pant's Landing to the National Eisteddfod platform. A few years later Hywel Price, Mostyn, accomplished an equally masterly feat, by winning the baritone competition in the National Eisteddfod. It is strange how the coal-dust affected the miner's chest, but not his voice.

The adjective 'numerous' is the one to use to refer to the local soloists and duetists. One partnership is sealed vividly in the memory. John y Farm and Dei Ben. Their stance on the stage was a feast to the eye. John, six-foot two inches tall and shoulders resembling a nine-foot bar, and Dei was not far behind. John was a baritone, and Dei Ben a deep bass, and if ever there was an embodiment of Deep Harmony it was Dei

Ben. They graced the North Wales concert halls and at times would cross the border to England. They gave a good account of themselves on the television programme, *Opportunity Knocks* when Hughie Green was compere.

Opportunity had previously knocked on John Herbert's door. During his youth, he travelled to Wrexham to have singing tuition from Dr Powell Edwards. The tutor immediately recognised the potential in the voice and informed John that he could help him to become a member of the Carl Rosa Opera Company, if he would leave the pit and follow a singing career. Some of John's mates in the pit at the time earned enough to buy a motor-bike . . . The B.S.A. won.

I also remember John for his bravery. In 1948 when he worked on the coal-face in the South of the Two-Yards, he heard a cry of 'help' from one of the nearby roads. He ran in the direction of the voice, and Twm Top the fireman close on his heels. Harold Davies, Gwespyr, the lad driving the pony at the time, followed quickly. John saw that a huge fall had occurred, and someone was buried almost completely beneath it. It was Abel Jones, unable to move a muscle. After making the roof a little safer, tons of rubble had to be cleared with pick and shovel to release the injured miner. After Abel was released Twm Top and Harold lifted him on John Herbert's back, and thus moved him a few yards to a safer place. Before they had gone twenty yards from the fall much more stone came down until the road was chock-a-block. Had they been a couple of minutes later they would all have been killed. The old man's condition was bad enough—both his legs were broken and his back severely injured. After tying his legs, he was put on a stretcher and John and Twm carried him to the surface along the fairly low South road, and up the very steep Two-Yard Sling. After they had seen their mate safe in the ambulance on his way to hospital, they were instructed by the management to go back to the coal-face, John to load, Twm to fire.

After Abel Jones recovered he remembered his rescuers, and he rewarded them. Recently I saw the beautiful silver

John Herbert Jones proudly displaying his silver tray.

tray which John received from him, and I decided that this incident had to be recorded for posterity. John did not receive a medal for his bravery in saving a comrade's life; the government of Great Britain awards medals for the killing of fellowmen.

The roof was not singing in the South of the Two-Yard Seam that day Abel Jones was injured, but through luck, or through Providence, there was a singer at hand to rescue him.

Through Luck . . . ? Through Providence . . . ? Through a lad's hand on the B.S.A. . . . ? Through God's hand on events . . . ? Through the call of the highway . . . ? Through God's mysterious Way . . . ?

Over to you, Professor Ap Nefydd.

These Were The Ponies Down Pit

No volume relating to the Point of Ayr can ignore the contribution of the ponies. It is true to say that every lump of coal that came up to the surface up to about 1960 was hauled part of the way by them. During the period I worked at the pit about forty of them were kept in underground stables. After coming to the pit as young ponies, most of them would hardly ever see much daylight again, if any. The 1921 and 1926 strikes were a blessing to the ponies at least, because they were brought to the surface, and they were allowed to prance on, and graze the lush pasture on the Point of Ayr fields. After lengthy years underground, when they were old, some were taken to the surface and let out to local small-holdings, to enjoy light-duties for the remainder of their days. The ponies were so well looked after, that even after they had been 'pensioned off' to the local small-holders the head-ostler would visit them periodically to see that they were well treated.

Occasionally, a horse would suffer a broken leg down pit, and he would then have to be humanely killed, then strapped to an empty timber lorry, and hauled to the surface with the coal tubs. This was a sad sight, of course, and although no one doffed his cap, there was plenty of sympathy there, and an occasional passing comment, 'Poor old Tyrffi.'

No coal mine has ever been an ideal place for a pony to work, but after saying that I hasten to add that the ponies at the Point of Ayr received every care and attention. Firstly, the pony had to be healthy, and a mare would not even be allowed to place the tip of her shoe underground. I don't know what some of our 'women libbers' would say about this. For my own part, if a woman in a factory or school is doing the same work as a man, then at all costs she should have equal pay and opportunities, but I have too much respect for her (and the mare) to say that the coal-mine is a suitable place for them. The law of the land, through the Coal Mines Act (1930), insisted that there were strict guidelines regarding the horses and the stables. 'All stables must be warm, and well ventilated'. The horses were not overworked, and details of their working hours and conditions were kept. Ostlers cared for them on the three shifts. It was they who gave them good beds, 'straw up to their bellies' as my father said, plenty of hay in the racks, bran in the manger, and a bucketful of clean water in each stall. It was they also who brushed and scraped them, and kept the stables very clean, through regular 'mucking out'.

It was they who sent for the farrier when one of the ponies was ill, and it was they who would also have an occasional nap on the sacks of feed in the warm stables. Only the jealous would blame them! Although sleeping underground was an offence, the ostlers were not the only ones to break an occasional section of the Coal-Mines Act. When I started in the pit the stables would be white-washed, but later they were coloured green for the ponies' sake.

When the drivers returned their ponies to the stables at the end of the shift, tethering them safely in the stall, the ostler would cast a quick glance over them to see that they were alright. Sometimes, if it was seen that there was excess sweat on one of them, the ostler would ensure that a stronger horse would be sent to that particular place the following day. Two full-time saddlers worked on the surface making collars, bridles, straps and traces as necessary. When one of the horses lost a shoe, one of the blacksmiths who worked on the surface would come down to shoe the pony, and he would receive a few extra mites in his wages each time he visited the stables, something like the travelling expenses claimed by councillors and clergymen, but a lot less!

When the horses first came down pit their coat was long, but in a few months, because of the warmth of the coal-face, and the daily treatment of the brush and scraper, the hair would become shorter, and the coat would shine like a gold sovereign. It was the regular nutritive food that they had that made them as round as apples, and the hard work they encountered made sure that the skin on the apple was not too thick.

One of the few things that remain in my memory from the Physics lessons in school is the saying, 'We see with the aid of light'. That is, we must have light before we can see anything. In those days in the pit electric lighting was very, very scarce, except in pit bottom and nearby levels. There were a few compressed air lamps, but in the remainder of the roads and in the coal-faces the only light was our little oil-lamps, to be followed at a later date by our cap-lamps. If one was to extinguish his lamp that would be a blackout indeed. Because there was no light, nothing could be seen, nothing, nothing, nothing. Ask our old Physics Master, Billy Boxer. Total blackness, and yet on occasion, the pony could, if he started out of the coal-face at the end of the shift before the driver, find his way to the stable through this utter underground darkness. I'm not teasing, but saying the truth.

Apart from this instinct, some of them could foresee, or sense, imminent danger. I have seen a horse, and have heard of many, stopping abruptly on a roadway, and refusing to move forward an inch. At first the reason would not be clear, but in no time there would be a roof-fall. Somehow or other the pony would sense the danger before the man. 'Dumb animals?'

'Not so dumb, lads bach.'

They were extremely intelligent and obedient animals, and in no time they would know when to turn back, where to step to one side, with only very few commands from the drivers. I saw Monty, a brown and white pony, perform a feat for his driver Glyn Fox that I have never seen another horse capable of doing. As the tubs were hauled out of the coal-face by the ponies the speed could be regulated by the driver placing scotches in the wheels. If a driver failed to get his scotches in, the tubs would accelerate, and would overturn, and the poor driver would have to fill them. This did not happen with Monty. When Glyn missed his scotches (which was not often!) he would shout at the pony, 'Hold 'em, Monty,' and the dear pony would stop in his tracks, haunch his back, and let the tubs bang against his back-side, avoiding a lot of work for Glyn. It is not my imagination at work, but my memory, and there are plenty still alive today who can vouch for these statements. It was only through kindness that the ponies were taught these feats.

Another unique partnership was the one between Horace Top and Churchill. As can be imagined, with such a name, this was not the easiest pony to handle, but through an occasional butty, a cake, or an apple from Horace, he became as easy to handle as a government back-bencher anxious for promotion.

One day Horace changed shifts, and another driver had to be found for Churchill. Unfortunately, during the shift the horse suffered an accident, and broke his leg. There was nothing to do but to send for Ifan Jones the head-ostler to bring the humane-killer to put the pony out of his pain. When Horace came to work on the afternoon shift he heard the sad news in the canteen. He broke down and cried and he went back home, too distressed to work that day.

'I've lost a good friend,' were his words to the canteen girls.

When miners walked together down pit they always did so in single-file, leaving a couple of yards space between each other. This was a safety measure of course. If a fall was to occur, far better for one to be beneath it than half a dozen. When the leader of the line would approach a low bar he would warn his successor by shouting 'Heads' and this warning would be repeated in turn, by everyone in the line, until the last had passed uninjured.

'Yu'll have to learn to stoop in more ways than one in this place my lad,' said my father, as I followed him in a rather low place on my first 'proper' shift in the pit. My father did not speak excessively, but there was plenty of commonsense and truth in what he said, and if I were to describe him in one sentence I would say that he was the opposite to a politician!

One day in 1949 half a dozen miners were walking to their work in the South of the Two-Yards. My father-in-law, Tommy Morris, led the line, although the pony Attlee was a couple of yards in front of him. The pony collided with a fairly low bar, and then, full of fright, he rushed backwards, crushing and fracturing Tommy Morris' leg. The break was severe, and my father-in-law spent many months in hospital. After recovering, and as he recalled the incident, he enjoyed saying, 'To think I've voted Labour all my life, and Attlee broke mi leg.' In case one believes that only politicians and generals were amongst the ponies, it would be better for me to record some of their names.

THE POINT OF AYR PONIES

Vacant the stables now, the lot,
   idle the brush, and ostler.
The embrocation is no more,
   and no one calls the farrier.

Once we had Blackburn, Twm and Marc,
   Pengwern and Cyrnol, Monty;
Tyrffi and Duke, the kicker Dic,
   once we had Sam and Batty.

Some bore a politician's name,
   Bevin and Churchill, Attlee,
but also a more honest breed,
   Collier and Saddler, Paddy.

The smallest one was Cherry Bach,
   Then Prins, and Pwnsh, and Dingo,
To boast the language of the pit
   one pony was named Cymro.

Dobbin, Llambert, and strangely, Wales.
   Turpin and also Cobbler,
Nelson and Bangor, Ben and Jim
   Jolly and Pinto, Trooper.

Now to the passbye no one turns,
   the flick'ring light is banished.
The coalface drivers curse no more,
   the horse-shoe sound has vanished.

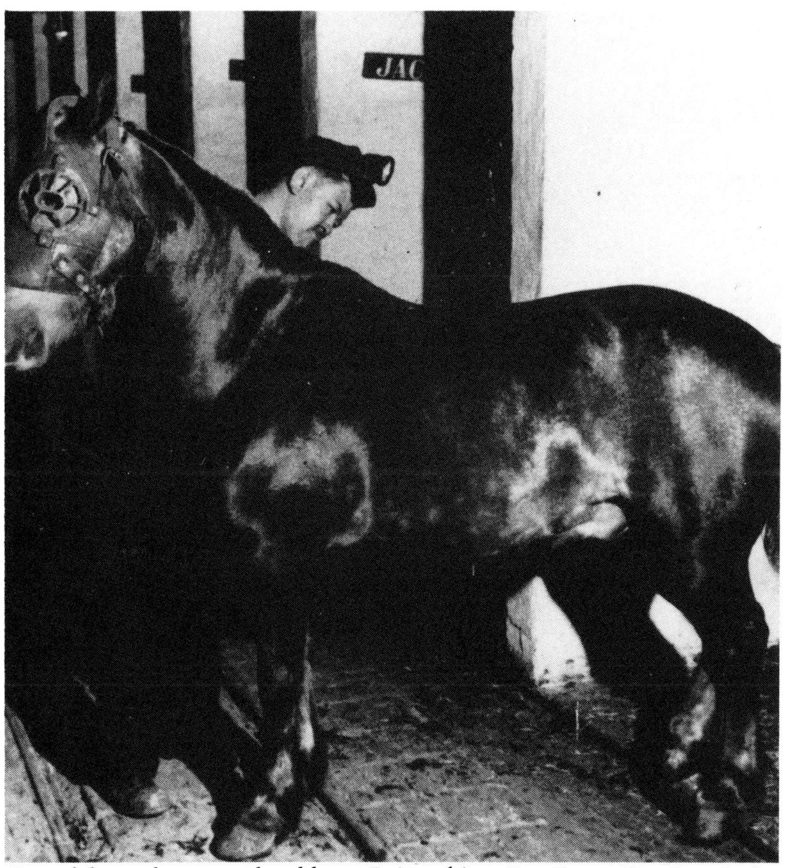

One of the under-ground stables at Point of Ayr. Ron Parry, Tan Lan, is the ostler.

'Corbet' and 'Twm' the day prior to the show in Rhyl.

It was the head-ostler's responsibility to name each horse, and in keeping with the Coal-Mines Acts to ascertain that the name was placed in bold letters on a piece of wood at the side of each stall, exactly as is seen on the office doors in the Shire Hall! The difference being, of course, that it was horses who were kept behind the name-plates in the pit. Sometimes a donkey loiters in the other place!

I will not tarry long considering the horses' names apart from noting the origin of two of them, namely Batty and Turpin. The last owner of the pit before nationalisation in January 1947 was Mr Batters of Tan-Lan Hall, and an abbreviated form of his name was given to the pony—poor dab. Turpin was christened after Randolph and not Dick. For a while the famous boxer trained at Gwrych Castle, Abergele, and married Gwyneth Price, a young Welsh girl from Acstyn. William Price (Will Cae Rhys) was one of the ostlers at Point of Ayr, and he was also Gwyneth's uncle, and because of the randiboo that surrounded the boxer, it was felt appropriate that the name should be brought to the pit by giving it to one of the ponies.

Two of the ponies had an unexpected 'shift for the king'

'Jolly' and 'Sam' the last two horses to work under-ground at Point of Ayr, 21st September 1968.
From left to right: Douglas Williams, 'Jolly', Elfed Hughes, Sam Blythin, 'Sam', Ernie Ormshaw, Cyril Parry.

about 1959. The Flint and Denbigh Show was held in Rhyl, and it was decided that year to have two of the pit-ponies there as an added attraction. The lot was cast for Twm and Corbett. Twm was a small, black horse, and because he had sustained an accident in the pit, his head was inclined a little to one side. It does not need a scholar to know what his nick-name would have been were he a man! It is difficult to describe Corbett's colour, as difficult as describing the colour of the Liberal-Democrats. I would describe it as a grey-ish-brown, but I'm sure the farmers have a proper name for this colour.

It was a week to remember for both horses. They were feasted like mayors, and did nothing, like mayors, with the people milling about them. Two ostlers from the pit cared for them, and I remember one of them entering into the spirit of the piece in earnest, by going there in corduroy breeches and leather leggins, if you please.

As a family we did not usually attend this 'Eisteddfod' but we knew that there was a special attraction for us there that year. At the time, Ennis was six years old, and she had listened to me relate countless stories to her about the ponies. Sometimes she would bake small cakes for the ponies, and she would learn from me, on my return, how much they had enjoyed her delicacies. During this period I found a horse-shoe belonging to one of the ponies in the coal-face, and I brought it home for Ennis. For obvious reasons it was not necessary to polish it, but my friend, Bill Manchester, made a brass stand for it, and Ennis treasured the horse-shoe through her life. The horse-shoe still remains in Meddfod. It was unmitigated joy for Ennis to visit Twm and Corbett in the show, time and time again, with a piece of cake and fistfuls of caramels. Caramels similar to the ones I used to buy in the little shop at the bottom of Gwespyr Rhiw years ago, when I went as a youngster with my father to fetch the coal.

A Lot in Which I'm Well Pleased

Loading three-score-and-ten tubs in the coal-face in one shift would be a good day's work, and according to the Good Book the same number of years is quite ample for anyone on this old earth.

The remainder, according to Auntie Mary, are 'the years of the ''If'' '. 'If through strength you reach four-score years.' Because I am old-fashioned enough to believe some of the things in the Bible, I realise that I shall not have to fill many more tubs before the Great Weighman calls me to account for all the tubs I sent to the surface during my life. He'll want to know why there was some dirt and stone-dust in some tubs, and why some of the tubs were little better than half full. In spite of that he'll notice a few tubs full of large lumps, another tub full of useful nuggets, and he'll remember the sweat and toil of many a bull-week and 'big-hooking'. He'll also notice that on the tubs I've filled towards the end of the shift that I've used his Son's tallies, and when he sees this special mark he'll forget everything about the 'shifts to the band' I had, and for every 'tip on road' in my experience. He'll also be willing to pay me a bonus. A completely unmerited bonus. Oh, lads bach, what a Weighman.

Weighman, check-weighman, banksman, onsetter, winding-engine man, loader, driver, dogi, fitter (including one hammer-fitter!), saddler, ostler, overman, fireman, railsetter, timberman, joiner, blacksmith, day-wage man, piecework man, these were the ones I worked with for fifteen years. With these I sweated, laughed, strained, joked, quarrelled, forgave, chatted, and argued. With these I lost my head, 'Be quiet, you Will-of-Two-halves.'

'Well you shut up, you Jim-Three-Quarters.'

'Boil your head, you old flannel.'

'You smoke your whiskers you old dish-cloth.'

'I'm glad that my father wasn't a crook.'

'You can shut your big gob up any day of the week, I've seen straighter double-turns than your father.'

But in the canteen over a cup of tea the words of the row would float away down the gutter. It was with these men that I shared the Woodbines and Capstans. In a word, everything was shared. The miner is unequalled for sharing. If he only has two drops of water left in his bottle, he'll willingly give one drop to his mate. When someone forgot to bring his snappin to work, there was no cause for concern; one would take the lid of his snappin-tin around the snappin-place, and would get a jam-butty from one, a marmalade butty from another, a spam, or corned-beef butty, or a piece of cake from another, until in the end he had a better snappin than everyone else, at least a more varied one. Many miners would spare the crusts around their butties, not wastefully, but for the pony-driver to have them for his pony.

Etiquette did not figure prominently in the snappin-place. When a man sat on a wooden tool-chest or on a track-sleeper he knew that he was nowhere near the Savoy; but I never once saw a miner who worked in the coal-face eating his snappin without removing his cap. 'In respect to the food' according to the old miners. 'In order that the fleas might have fresh air' according to Dan Burum. I never heard anyone use the strict poetic metre (Cynghanedd) in the pit, except maybe, unknown to them, sometimes by swearing, 'Oh, hell-fire, you silly fools.' Swearing is a phenomenon associated with facing danger, I believe. It was easy to have pure lips when one was half asleep listening to the 'State of the Cause' in the Calvinistic Monthly Meeting, not so easy when the tubs were off-rails, and no one near to give a hand to lift them back, and the impatient loaders knocking the airpipe with a pick as a sign that they were waiting for more empty tubs to fill. No, I believe that an occasional little curse at times like this was as useful as a wooden crow-bar when one of those was not available. But woe betide the one that swore in the cage. When someone slipped into that transgression he would be reminded abruptly by one of the others, 'Remember, lads bach, where we are, remember that we're hanging on a thread.'

With great respect and admiration for these workers I wrote:

DUST

Put down that shovel, Bob, and come this way,
to have a swig and some jam-butties too.
We'll let the dust, and smoke-fumes drift away,
I need to ease my back, and so do you.
Do you recall the miners long ago
who tamed the Five-Yards and the stubborn Cant?
Rhobat Tan-Capel, Ned y Garth and Joe.
M'redydd and Tommy'r Moelfryn, Twm y Pant.
Peace to their dust! True mates that would not shirk,
the life they led was never one of ease.
Craftsmen, who got a pittance for their work,
but the Big Book supplied God's cheques for these.
Hard-workers all, now take a rest they must,
salt of the earth, they sanctify the dust.

There were a few exceptions of course who carried fowls and farm-butter to the bosses, but most of them were truly the salt of the earth. It has been my privilege to know them, to work with them, and to be accepted as one of them.

The Cockney boasts because he can say that he was born within the sound of Bow Bells; I also square my shoulders because I have spent all my life within the sound of the Point of Ayr hooter.

A few months ago I met Wil Washars sitting on a seat in the entrance to the Kwik Save Supermarket, Prestatyn. We had not met for years. Will had left the pit a little before reaching middle-age, although his father and brothers remained in the pit all their lives. Will's sweetheart lived in Colwyn Bay, and after quite a long courtship they were married. Living tally was not fashionable in those days! They set up home in

Colwyn Bay, and therefore, because of the long travelling distance Will had to bid farewell to the pit.

'Hei, wa' d'yh think? Will Washars is working wi' the rag-an-bone man.'

That was the Stop-Press I received one morning as I collected my lamp from the lamproom. The sun and the moon shone brightly behind Donald Relwê's eyes as he gave me this scoop.

'Don't talk rubbish.'

'I'm not talking rubbish. Rubbish is the stuff that the old Washars is collecting, yes, I'll be damned, his brother Pit told me this morning.'

'Shiver-me-pit-props! Love makes a man do strange things at times!'

Will a rag-an-bone-man. That was what loaded the conversation on the way to the coal-face, snappin-time, and then on the way out to pit-bottom. He would have created less of a stir had he gone to the moon.

It is said that old Jo—Strêt-ffês only laughed twice in his life; once at a quarter-past three on August 29th, 1933, and also when he heard about Will Washars and his new job.

Will, with his thundervoice, turning the supermarket entrance into Ifor Tan Rhiw's passbye:

'Hello, my old timer, give me y'uh paw.'

'Hai, me old Wash?'

'Put y'ur arse on this seat, me old Ein, for us to have a ''sight pause''.'

Before the introduction of the cap-lamps to the pit, when the miners were wholly dependent on the little flame-lamps, they would pause for two or three minutes in pit-bottom for their eyes to become accustomed to the darkness, and then they would proceed to their place of work.

'How y'uh keepin, Will?'

'Great, considering that I'm three-score-and-thirteen this year,' said Will, using the form of counting used in the pit to count the tubs leaving the coal-face. To show that I had not forgotten my roots either I said that I was three-score-and-four.

'Time's goin' mate. You nor I will have to buy many more tallies from Ned Wedyn to put on the tubs of our days.'

'Will, y'u'r talkin' like a poet.'

'Don't mention those things to me. Don't yuh think its enough for you to be a poet, without me bellowing about things no one understands?'

'D'y'uh remember . . .?'

'D'y'uh remember . . .?'

'D'y'uh . . . hey, look, for goodness sake, who's over there Rudolph y Gors. I haven't seen him since we were dropping the tops in the Old Durbog, years ago. Come here, Rudolph Bach for us to see yuh.'

'How go, lads bach?'

'D'y'uh remember this Bevin Boy? Sit on the inbye-side, Rud. That's it, the three musketeers.'

'Y'uh'r not alt'rin' at all, Will.'

'D'y'uh remember . . .?'

'D'y'uh remember . . .?'

'Does this Kwik Save felluh sell chewing-baccy, tell me?'

'Did yuh chew baccy on the scrap-lorry, Will?'

Ha! Ha! Ha!

'Knock twelve on that pipe, Rud.'

'Home James, lads bach. So long!'

Four generations of the same family who worked at Point of Ayr.

(a) John Morris Yr Ardd Ddu, who was crippled in a mining accident at the Point of Ayr around the turn of the century.

(b) The miners receiving their first holidays-with-pay packets in 1939. Tommy Morris amongst them.

(c) John Morris (Tommy's son) looking David Hunt straight in the eye, when he was a high-ranking Coal-Board Official. For that matter, John can always look at everyone eyeball-to-eyeball.

(d) Nefyn Morris (John's son) and his mate Andy Hutchinson at work.

TO HELP WITH SOME OF THE TERMS

Bach—Literally small. When used after a person's name it has tones of endearment.

banksman—the man at the pit-head who gives signals to control the movement of the cage.

chain-clip—a steep coal-face, where in addition to scotches being used to slow the speed of the tubs, a chain was also dragged behind the tubs.

check-weighman—a person who sat next to the weighman on surface checking the weighing and recording of the tubs. His wages were paid by the coal-cutters and loaders and not by the management. He was a nuisance to the management, but a friend to the workers.

dogi—general helper.

hammer-fitter—a useless fitter, one who could only use a hammer.

hooking shift—production shift.

inbye—inwards.

iron-man—a coal cutting machine.

jig—an incline where the only power used to move the tubs of coal was gravitation.

knock twelve on the pipe—to knock the compressed-air pipe with a pick or scotch to denote the end of the shift. Six knocks indicated snappin-time.

onsetter—the man at pit-bottom who gives signals to control the movement of the cage.

outbye—outward.

passbye—where the single track becomes double track for about twenty yards, and full tubs run on one track, and empty tubs on the other.

pit-head plates—large steel platform at the pit-head.

riding the tail-chain—in order to manoeuvre the coal-tubs around difficult bends in the coal-face the pony driver would jump on the horse's tail-chain and use the pressure of his legs and arms to guide the tubs. This was contrary to the Coal-Mines-Acts and was not done when the H.M. Inspectors were around.

skilled shilling—after learning many skills a miner could, at the discretion of the management, be awarded an extra shilling a day. By now the amount seems trivial, but when a little overtime was worked this would amount to seven-and-six a week. That was the amount of my rent at the time.

sling—incline.

snappin—the food eaten in work.

sprag—a wooden support for upholding the sides in the coal-face.

tallies (metal)—each miner would hand his numbered tally (disc) to the banksman as he went into the cage. The tallies were put on a board at pit-head, and collected by each miner on his return to surface. Any tallies left on the board at the end of the shift indicated that a miner was either working overtime, or was missing, in which case enquiries would be made.

tallies (leather)—each set of four men working in the coal-faces would have a number. This number was cut on a small piece of leather and tied by thick string to each tub leaving the coal-face, so that it could be recorded on surface by the weighman, and the appropriate set paid accordingly. The tallies were made by Ned Griffiths (Ned Wedyn) who collected pieces of leather scraps from the Ffynnongroyw cobbler. He charged half-a-crown a score for the tallies, and fair play for Ned he was always generous in his counting.

waiting time—if the loaders at the coal-face were kept waiting a long time for empty tubs to fill they were paid a small allowance.

water money—in coal-faces that were excessively wet the men would be paid a small allowance. Yes, allowances were always small.